Lecture Notes in Mathematics

Edited by A. Dold and B. Eckmann

1246

Hodge Theory

Proceedings of the U.S.-Spain Workshop
held in Sant Cugat (Barcelona), Spain
June 24–30, 1985

Edited by E. Cattani, F. Guillén, A. Kaplan and F. Puerta

Springer-Verlag
Berlin Heidelberg New York London Paris Tokyo

Editors

Eduardo Cattani
Aroldo Kaplan
Department of Mathematics, University of Massachusetts
Amherst, Massachusetts 01003, USA

Francisco Guillén
Fernando Puerta
Departament de Matemàtiques ETSEIB
Universitat Politècnica de Catalunya
Avenida Diagonal 647, 08028 Barcelona, Spain

Mathematics Subject Classification (1980): Primary: 14C30, 32J25
Secondary: 14B05, 14B15, 14D05, 14F40, 14J15, 32C38, 32G13, 32G20

ISBN 3-540-17743-4 Springer-Verlag Berlin Heidelberg New York
ISBN 0-387-17743-4 Springer-Verlag New York Berlin Heidelberg

Printed in Germany

Printing and binding: Druckhaus Beltz, Hemsbach/Bergstr.
2146/3140-543210

Introduction

This volume contains most of the papers presented at the 1985 U.S.-Spain workshop on Hodge Theory, which took place in June of that year in St. Cugat del Vallés (Barcelona), under the auspices of the Joint Committee for Scientific and Technological Cooperation between the two countries.

The methods of harmonic theory introduced by Riemann in the study of algebraic curves and extended by Hodge to higher dimensions, yield important consequences for the structure of smooth projective varieties; prototypical examples are the decomposition of the complex cohomology groups of such varieties according to Hodge type, and the Lefschetz decomposition in terms of primitive cohomology classes. Since Hodge, these ideas have undergone several generalizations of great interest in algebraic geometry. This volume focuses on those concerning singular and open varieties, families, and the rational homotopy of varieties, which have been developed during the past 20 years.

The structure present in the local system of cohomology defined by a family of smooth varieties was encoded by Griffiths in the notion of a variation of (polarized) Hodge structure[1]. Some of the related questions explored in these proceedings include: a geometric realization of certain maximal variations of Hodge structure; an asymptotic description of a variation degenerating at a divisor with normal crossings; an explicit construction, for some particular cases, of the equivalence

---

[1] cf. the survey papers:
P. Griffiths: "Periods of integrals on algebraic manifolds: summary of main results and discussion of open problems", Bull. Amer. Math. Soc. 76, 228-296 (1970).
P. Griffiths and W. Schmid: "Recent developments in Hodge Theory: a discussion of techniques and results" , Proc. of the International Colloquium on Discrete Subgroups of Lie Groups and Applications to Moduli, Bombay 1973, Oxford University Press, 1975, 31-127.

between unipotent variations of mixed Hodge structure and mixed Hodge theoretic representations of the fundamental group, and a study of the higher Albanese manifolds that play a role in the classification of unipotent variations of mixed Hodge structure.

In "Théorie de Hodge, I, II and III"[2], Pierre Deligne introduced the notion of a mixed Hodge structure and proved the existence and functoriality of such a structure in the (ordinary) cohomology groups of singular or open varieties over **C**. Two very fruitful approaches to the study of mixed Hodge theory are presented in this volume: the use of cubical hyperresolutions and the notion of iterated integrals. They are applied to the construction of mixed Hodge structures on the rational homotopy of algebraic varieties, on the cohomology of a link, on the local cohomology of an analytic space in the neighborhood of a compact algebraic subvariety, and on the vanishing cycles. An ℓ-adic description of the weight filtration on the cohomology of a link is also given, as well as a result showing that mixed Hodge complexes are preserved by truncations.

A third subject discussed here is that of the L_2- realization of Intersection cohomology and its implication: the existence of pure Hodge structures on these groups. The cases of singular varieties with constant coefficients and of degenerating coefficients on a smooth base are both treated extensively. Finally, connections with arithmetical questions and an update on certain aspects of the Hodge conjecture are included as well.

The limitations inherent in the format of the workshop made it impossible to include other important aspects of modern Hodge theory. Among these we should mention the applications to the study of singularities, algebraic cycles, the theory of infinitesimal variations of Hodge structure, and Torelli-type

---

2 Théorie de Hodge: I, Actes du Congrès International des Mathematiciens, Nice 1970; II, III, Publ. Math. IHES, 40 (1972), 5-57 ; 44 (1974), 5-78.

problems. Some of these questions are extensively discussed in "Topics in Transcendental Algebraic Geometry", P. Griffiths, ed., Annals of Math. Studies, Princeton U. Press (1984). We also refer to this volume as an introduction to the basic concepts and methods.

We are very grateful to all the participants who made this a succesful conference and, very specially, the contributors to this volume. Thanks are also due to Dr. Harold Stolberg of the National Science Foundation for his support.

October 1986.

E. Cattani
F. Guillén
A. Kaplan
F. Puerta

Table of Contents

Shimura Varieties of Weight Two Hodge Structures

by

James A. Carlson and Carlos Simpson

1. Introduction

Consider a variation of Hodge structure of weight two,

$$v : Y \longrightarrow \Gamma\backslash D .$$

Griffiths transversality [6,7] asserts that vectors tangent to the image of v lie in the so-called horizontal tangent bundle $T_h(\Gamma\backslash D)$, which has fiber dimension $h^{2,0}h^{1,1}$, so that

$$\text{(1.1)} \qquad \dim v(Y) \leq h^{2,0}\, h^{1,1} .$$

When $h^{2,0} > 1$ the horizontal vectors define a non-integrable distribution in the holomorphic tangent bundle. But v(D) is an integral submanifold for T_h, so that additional restrictions hold, namely

$$\text{(1.2)} \qquad \dim v(Y) \leq \frac{1}{2} h^{2,0} h^{1,1} .$$

In the category of abstract variations of Hodge structure this bound is best possible [2]. The purpose of this note is to show that the bound is also best possible in the category of geometric variations of Hodge structure.

To make the notion "geometric" precise, we follow [8]: Let AF be the category whose objects are smooth families of algebraic varieties, $X/Y = [\pi : X \longrightarrow Y]$, and whose morphisms are given by commutative diagrams of morphisms of varieties. Let $GVHS_0$ be the category whose objects are the variations of Hodge structure $R^k\pi_*\mathbf{C}$ defined by objects of AF and whose morphisms are those induced from AF. Let GVHS be the smallest abelian category which contains $GVHS_0$ and which is closed under formation of tensor products and duals.

The weight two variations to be constructed satisfy a symmetry condition: their Hodge structures admit an automorphism J such that (i) $J^2 = -1$, (ii) $S(Jx,Jy) = S(x,y)$, where S is the polarizing form, and (iii) $J|H^{2,0} = +i$. We shall call such an object a J-Hodge structure. J-Hodge structures are classified by a space with a transitive U(p,q) action with isotropy subgroup U(p)×U(q), so that there is a natural identification with the generalized unit ball,

$$(1.4) \qquad B_{pq} = \{ \text{ complex } p\times q \text{ matrices } \mid {}^t\bar{Z}Z < I \} .$$

(See [9], p. 527, domains of type A III). Moreover, the natural family of J-Hodge structures over B_{pq} actually defines a variation of Hodge structures with $h^{2,0} = p$ and $h^{1,1} = 2q$. Since $\dim B_{pq} = pq = (1/2)h^{2,0}h^{1,1}$, the bound of (1.2) is satisfied. We shall also refer to variations of J-Hodge structures as unitary.

To show that unitary structures arise from geometry, we consider unitary weight one structures: structures admitting an automorphism J as above, but where the dimension of the +i-eigenspace of $J|L^{1,0}$ (call it p) is arbitrary. Let E be the Hodge structure on H^1 of the elliptic curve with period lattice generated by 1 and i. Then E carries a natural J-structure, induced by multiplication by i on the underlying curve. Let L⊗E be the tensor product of these two unitary structures, and observe that J⊗J is a morphism with eigenvalues ±1. Define the Prym structure associated to L to be the −1 eigenspace of J⊗J:

$$(1.5) \qquad P(L) = \text{kernel } [J\otimes J + 1 : L\otimes E \longrightarrow L\otimes E] .$$

This structure is unitary with respect to J⊗1. The main result is then:

<u>Theorem</u> (1.6) If (H,J) is a J-Hodge structure of weight two, then there exists a unitary weight one structure (L,J) such that

$$(P(L),\, J\otimes 1) \cong (L,J).$$

<u>Corollary</u> (1.7) Every J-Hodge structure of weight two arises from geometry.

2. Unitary variations of Hodge structure

To describe the variations which realize the bounds we fix the following terminology. A <u>lattice</u> is a pair (Λ,S) consisting of a torsion-free abelian group Λ and a non-degenerate integer-valued bilinear form S on Λ. A <u>complex structure</u> on a lattice is an endomorphism J of Λ satisfying (i) $S(Jx,Jy) = S(x,y)$ for all x and y, and (ii) $J^2 = -1$. We shall also call a lattice with a complex structure a <u>J-lattice</u>. A morphism of J-lattices $\phi : \Lambda_1 \longrightarrow \Lambda_2$ is a group homomorphism satisfying (i) $J_2\phi = \phi J_1$, (ii) $S_2(\phi x,\phi y) = S_1(x,y)$ if the two forms are either both symmetric or both antisymmetric, (iii) $S_2(\phi x,\phi y) = S_1(Jx,y)$ otherwise. Note that in general the bilinear form

$$(2.1) \qquad {}^{J}S(x,y) = S(Jx,y)$$

is symmetric if J is antisymmetric (and viceversa). A <u>J-polarized Hodge structure</u> consists of (i) a polarized Hodge structure and (ii) a J-structure on the underlying lattice, where J is a morphism of Hodge structures.

J-lattices exist. For a trivial example, take a lattice (L,S), consider the doubled lattice $(L,S_+) = (L\oplus L,\ S\oplus S)$, then set $J(x,y) = (-\phi^{-1}(y),\ \phi(x))$, where $\phi : L \longrightarrow L$ is an isometry. The E_8 lattice gives a nontrivial (indecomposable) example. To see this, view E_8 in Q^8 with root system Δ given in the standard basis by

$$\pm e_i \pm e_j \quad (1 \le j < i \le 8) \; ;$$
$$(1/2)(\Sigma\, \varepsilon_i e_i), \quad \text{where } \varepsilon_i = \pm 1, \quad \Pi\varepsilon_i = -1 \; .$$

Let J be the linear transformation of Q^8 defined by

$$J(x_1, \dots, x_4, x_5, \dots, x_8) = (-x_5, \dots, -x_8, x_1, \dots, x_4)$$

and observe that $J(\Delta) = \Delta$, so that J restricts to an automorphism of the E_8 lattice of the required type. One may construct pairs (S,J) with associated hermitian form of prescribed signature (p,q), as in the matrix example below:

$$S_+ = \begin{bmatrix} I_{2p} & 0 \\ 0 & -I_{2q} \end{bmatrix}, \quad J = \begin{bmatrix} J_p & 0 \\ 0 & J_q \end{bmatrix}, \quad \text{where} \quad J_p = \begin{bmatrix} 0 & -I_p \\ I_p & 0 \end{bmatrix} .$$

Let $D = D(\Lambda, S, n, \underline{h})$ be the classifying space for S-polarized Hodge structures of weight n with Hodge numbers $h^{p,q} = \underline{h}^{p,q}$. (We suppose here that $h^{p,q} = 0$ for $p < 0$.) According to [7], D is a homogeneous space for a real Lie group $G(D) = SO(\Lambda_{\mathbf{R}}, S) \cong SO(a,b,\mathbf{R})$, where in general $\Lambda_K = \Lambda \otimes K$. The discrete subgroup $\Gamma = SO(\Lambda_{\mathbf{Z}}, S)$ acts on D to give an analytic quotient space $\Gamma \backslash D$ which is quasiprojective in the Hermitian symmetric case [1].

Fix a complex structure J on (Λ, S), fix integers $\underline{e}^{p,q}$, and let $D(J, \underline{e})$ be the set J-polarized Hodge structures in D for which the dimension of the +i-eigenspace of $J|H^{p,q}$ is $\underline{e}^{p,q}$. Thus $D(J, \underline{e})$ is a homogeneous space for the real Lie group $G(D,J) = SO(\Lambda_{\mathbf{R}}, S, J)$ consisting of elements of $SO(\Lambda_R, S)$ which commute with J. To identify this group consider the decomposition $\Lambda_C = \Lambda_+ \oplus \Lambda_-$ into the $\pm i$-eigenspaces of J, an orthogonal decomposition relative to the hermitian form $h(x,y) = i^n S(x,y)$ associated to S. Then

(2.2) the restriction map $g \longrightarrow g|\Lambda_+$ defines an isomorphism of $G(D,J)$ with the unitary group of h on Λ_+ --- an indefinite unitary group of type $U(c, d)$.

Once again, analytic quotients $\Gamma(J)\backslash D(J,\underline{e})$ are defined, where $\Gamma(J)$ denotes the (arithmetic) subgroup of $G(D,J)$ which preserves the lattice. When the isotropy subgroup of a reference structure F^*, written $G(D,J,F^*)$, is maximal compact, i.e., of the form $U(c)\times U(d)$, then $D(J)$ is hermitian symmetric and the discrete quotient is quasiprojective.

We now study the construction $D(J)$ in the weight one case. Let L_Z denote the underlying lattice, and let $L = (L_Z, S, F^*)$ denote an S-polarized Hodge structure of weight one and genus g: $\dim L^{1,0} = g$. Let p denote the dimension of the $+i$-eigenspace of J on L, and write H_g for D, $H_g(J,p)$ for $D(J,\underline{e})$, where H_g is the Siegel space of genus g.

<u>Theorem</u> (2.3) Let $H_g(J,p)$ be a weight one unitary space. Then

(a) the isotropy group of a reference structure is isomorphic to $U(p)\times U(q)$, where $p+q = g$,

(b) $H_g(J,p)$ is hermitian symmetric,

(c) $H_g(J,p)$ is a complex submanifold of H_g.

<u>Proof</u>. Because J is a morphism of Hodge structure, one has a decomposition $L_+ = L_+^{1,0} \oplus L_+^{0,1}$. Let $h(x,y) = iS(x,y)$ be the Hermitian form associated to S. The Riemann bilinear relations imply that h is

positive on $L_+^{1,0}$ and negative on $L_+^{0,1}$, hence is of signature (p,q). Since an element $g \in G(H_g,J)$ preserves the type decomposition on L_+, it maps to an element of $U(p)\times U(q)$; the map $G(H_g,J,L) \longrightarrow U(p)\times U(q)$ is then easily verified to be an isomorphism.

To see that $H_g(J,p)$ is Hermitian symmetric, observe that the associated special unitary group, $SU(p,q)$ also acts transitively, with isotropy group $S(U(p)\times U(q))$. This latter group is maximal compact with one-dimensional center, as required.

To show that $H_g(J,p)$ is a complex submanifold of H_g one may use the fact that the imbedding is defined by an imbedding of Lie groups in which the (one-dimensional) centers of the isotropy groups correspond:

(2.4.a) $\quad H_g \cong Sp(g)/U(g)$

(2.4.b) $\quad H_g(J,p) \cong SU(p,q)/S(U(p)\times U(q))$

where $g = p+q$. One may also give a direct argument. The complex structure on H_g is that given as an open subset of a subvariety of a Grassmanian:

(2.5) $$H_g \subset \check{H}_g = \{F \subset L_C \mid \dim F = g \text{ and } S|F = 0\},$$

where $F = L^{1,0}$. The locus $H_g(J,p)$ is defined by the Schubert conditions

(2.6) $$\dim(F \cap L_+) = p \quad \text{and} \quad \dim(F \cap L_-) = q,$$

and so is a complex submanifold.

The discrete quotient $Y = \Gamma(J)\backslash H_g(J,p)$ is quasiprojective, and is an instance of a Shimura variety [10,11,12,13]. Replacing Γ by a subgroup of finite index which acts without fixed points, one may construct a family of Abelian varieties A/Y which admits a globally defined endomorphism J of square -1. Thus Y classifies "Abelian varieties with additional structure" (i.e. J).

We study next the weight two case:

Theorem (2.7) Let D(J) be a weight two unitary space with $h^{2,0} = p$, $h^{1,1} = 2q$, and $J|H^{2,0} = +i$. Then

(a) the isotropy group of a reference structure is isomorphic to $U(p)\times U(q)$,

(b) D(J) is Hermitian symmetric,

(c) D(J) is a complex submanifold of D,

(d) D(J) is tangent to the horizontal distribution of D,

(e) D(J) defines a variation of Hodge structure of maximal dimension.

Proof: Denote the underlying lattice by $H_{\mathbf{Z}}$ and fix a reference Hodge filtration $F^* \in D(J)$. Since J preserves $H^{1,1}$, there is an eigenspace decomposition $H^{1,1} = H_+^{1,1} \oplus H_-^{1,1}$. Let $H_{\mathbf{C}} = H_+ \oplus H_-$ be the eigenspace decomposition on the complexification of the lattice, and let $h(x,y) = -S(x,\bar{y})$ be the Hermitian form associated to S. By the Hodge-Riemann bilinear relations, $h|H_+$ has signature (p,q). An element $g \in G(J,F^*)$ preserves the refined type decomposition

$$(2.8) \qquad H_{\mathbf{C}} = [H^{2,0} \oplus H_+^{1,1}] \oplus [H_-^{1,1} \oplus H^{0,2}],$$

where the first term in brackets is H_+ and the second term is H_-. Therefore $g|H_+$ lies in $U(p)\times U(q)$. This proves (a), and (b) follows, since the isotropy group of the associated special unitary group is maximal compact with one-dimensional center.

To prove holomorphicity, note that by (2.8) D(J) is (once again), defined by a Schubert condition: $F^2 \subset H_+$.

To prove horizontality, we use the dual Schubert condition $H_+ = [H^{2,0} \oplus H_+^{1,1}] \subset F^1$. Consider therefore a holomorphic curve of filtrations $F^*(t)$, and let $\omega(t) \in F^2(t)$ define a holomorphic section of the Hodge bundle F^2. Then $J\omega(t) = +i\omega(t)$. Differentiation yields $Jd\omega/dt = +id\omega/dt$, so that $Jd\omega/dt \in H_+ \subset F^1(t)$. Thus $dF^2/dt \subset F^1$, as required.

The last assertion follows from (1.2), since the dimension of $U(p,q)/U(p)\times(U(q)$ is $pq = (1/2)h^{2,0}h^{1,1}$.

3. The Prym construction

In this section we prove theorem (1.6). The first step is to define canonical operators which change the weight of a J-Hodge structure. Suppose given a J-Hodge structure of weight two, $H = (H_{\mathbb{Z}}, S, F^*, J)$. Define a new filtration ${}^JF^*$ by

$$(3.1) \qquad {}^JF^1 = H^{2,0} \oplus H_-^{1,1} \qquad {}^JF^0 = H_{\mathbb{C}} .$$

Then $(H_{\mathbb{Z}}, {}^JS, {}^JF^*, J)$ is a J-polarized Hodge structure of weight one. One thinks of this as follows: The +i eigenspace of J has formal type (1,0), while the -i eigenspace has formal type (0,1). Subtract formal type from actual type to get the type of the "J-twisted" Hodge structure. According to this rule, ${}^JF^1$ has type (1,0).

Given a J-polarized Hodge structure of weight one, $H = (H_{\mathbb{Z}}, S, F^*, J)$, define a new filtration by

$$(3.2) \qquad {}^JF^2 = H_+^{1,0} \qquad {}^JF^1 = H_+^{1,0} \oplus H_-^{1,0} \oplus H_+^{0,1} \qquad {}^JF^0 = H_{\mathbb{C}}.$$

The correct definitions are arrived at by the same rule, except that one adds actual and formal types to get the new type. The new object $(H_{\mathbf{Z}}, {}^{J}S, {}^{J}F^{*}, J)$ is a J-polarized Hodge structure of weight two. Note that $^{J}({}^{J}H) = H$, so that the operation $H \longrightarrow {}^{J}H$ defines a natural isomorphism between the categories of J-Hodge structures of weights one and two, respectively.

The next (and final) step is the following:

__Lemma__ (3.3) For every J-polarized Hodge structure of weight 2 there is a canonical isomorphism

$$\phi : H \longrightarrow P({}^{J}H) .$$

__Proof__: Let P_{+} and P_{-} be the projections of $H_{\mathbf{C}}$ onto the $+i$ and $-i$ eigenspaces of J. Let $E = (E_{\mathbf{Z}}, S_{E}, F^{*}, J)$ be the J-Hodge structure of the elliptic curve with period ratio i. Let (e, e') be a symplectic basis for $E_{\mathbf{Z}}$, and let $\omega = e + ie'$ generate $E^{1,0}$. Define a linear transformation

$$(3.4) \qquad \phi : H_{\mathbf{C}} \longrightarrow H_{\mathbf{C}} \otimes E_{\mathbf{C}}$$

by the formula

$$(3.5) \qquad \phi(v) = P_{+}(v) \otimes \omega + P_{-}(v) \otimes \bar{\omega}.$$

Because $J \otimes J(\phi(v)) = -\phi(v)$, the image of ϕ lies in the complex vector space of $P({}^{J}H)$. Since the terms of $\phi(v)$ lie in distinct eigenspaces of

$J\otimes 1$, $\phi(v)$ vanishes if and only if both $P_+(v)$ and $P_-(v)$ vanish, i.e., if and only if $v = 0$. Consequently

(3.6) $\phi : H \longrightarrow P({}^JH)$ is an isomorphism of vector spaces.

To see that ϕ preserves the integral structure, substitute the relations $P_\pm(v) = (1/2)(J \pm i)$ into the definition of $\phi(v)$ to obtain

(3.7) $$\phi(v) = Jv\otimes e - v\otimes e',$$

and observe that J preserves the integral structures.

We verify that ϕ preserves polarizations, where $L\otimes E$ carries, up to scale factor, the canonical polarization of a tensor product,

(3.8) $$S(x\otimes y,\ x'\otimes y') = (-1/2)\,{}^JS(x,x')S_E(y,y') .$$

To this end substitute the definition of $\phi(v)$ and $\phi(v')$ in the definition of S and use the fact that the +i and -i eigenspaces of J are S_+-orthogonal to obtain

$$\begin{aligned} S(\phi(v),\phi(v')) &= (-1/2)[{}^JS(P_+v,\ P_-v')S_E(\omega,\bar{\omega}) + {}^JS(P_-v,P_+v')S_E(\omega,\bar{\omega})] \\ &= [\ S(P_+v,P_-v') + S(P_-v,P_+v')\ ] \\ &= S(v,v'). \end{aligned}$$

Since ϕ preserves the integral structures, it is defined over the real numbers. Thus, to show that ϕ preserves Hodge filtration, it suffices to show that $\phi(H^{2,0}) \subset (H\otimes E)^{2,0}$. If v is in $H^{2,0}$ then $P_+(v) = v$ and $P_-(v) = 0$, so that $\phi(v) = v\otimes\omega \in (H\otimes E)^{2,0}$, as required.

To summarize, we have:

Corollary (3.9) The natural variation of Hodge structure over $\Gamma(J)\backslash D(J)$ arises from geometry.

4. Higher weight

To discuss the higher weight situation we recall the notion of an infinitesimal variation of Hodge structures (IVHS [3, 4]). Fix a Hodge structure H of weight n, let G^* be the graded object associated to the Hodge filtration, so that $G^p = F^p/F^{p-1} \cong H^{p,q}$, and observe that the polarization on H descends to a bilinear form on G^*. Let End^* be the graded Lie algebra of endomorphisms of G^*, and let E^* be the subalgebra of antisymmetric endomorphisms. Then E^{-1} is naturally identified with the horizontal tangent subspace of the holomorphic tangent space to D at H, where D is the classifying space associated to H. An infinitesimal variation of H is a subspace T of E^{-1} which satisfies the integrability condition $[T, T] = 0$. The tangent spaces of variations of Hodge structure define infinitesimal variations.

Call an infinitesimal variation nondegenerate if the natural map

$$(4.1) \qquad T \times G^p \longrightarrow G^{p-1}$$

is surjective for $m < p \leq t$, where m is the middle Hodge index, $m = [\text{weight}/2]$, and where $t = \max\{p \mid G^p \neq 0\}$ is the top Hodge index. Then one has the following dimension estimate:

Theorem (4.2) : Let (H,T) be a nondegenerate infinitesimal variation of weight n. Then

$$\dim T \leq h^{t,n-t}\, h^{t-1,n-t+1} .$$

Proof : Use the proof of Theorem 4.1 of [2] and note that the slightly weakened nondegeneracy hypothesis used here suffices because of antisymmetry. (Erratum: the first factor ($h^{t,n-t}$) is omitted both in the statement and in the proof of Theorem 4.1 [2])

We now construct a geometric variation of Hodge structures of weight n which satisfies the bound of the preceding theorem. To begin, consider the variation of weight one structures L/Y defined in Theorem 2.3, let E/Y be the constant weight one variation of genus one with period ratio i, and let J stand for the canonical automorphism of square -1 for L/Y and for E/Y. Form the weight n variation $H = L \otimes E^{\otimes(n-1)}$. On H consider the endomorphisms

$$(4.3) \qquad \pi_{rs} = J_r J_s + \mathrm{id} ,$$

where J_r acts on the tensor product by J in the r-th factor and by the identity in the remaining factors. Define a generalized Prym structure of weight n by

$$(4.4) \qquad P(L,n) = \bigcap_{1 \leq r < s \leq n} \ker(\pi_{rs})$$

For a vector to be in P(L, n), either J_r must act with eigenvalue +i for all r, or J_r must act with eigenvalue -i for all r. Consequently all Hodge components of P are zero except for those below :

$$
(4.5)\qquad
\begin{aligned}
P^{n,0} &= (L_+)^{1,0} \otimes (E^{1,0})^{\otimes(n-1)}\\
P^{n-1,1} &= (L_+)^{0,1} \otimes (E^{1,0})^{\otimes(n-1)}\\
P^{1,n-1} &= (L_-)^{1,0} \otimes (E^{0,1})^{\otimes(n-1)}\\
P^{0,n} &= (L_-)^{0,1} \otimes (E^{0,1})^{\otimes(n-1)}
\end{aligned}
$$

One checks that the natural map from T to $\mathrm{Hom}(P^{n,0}, P^{n-1,1})$ is an isomorphism, so that the variation P/Y is nondegenerate and satisfies the bound of theorem (4.2).

Using the generalized Prym construction it is possible to construct geometric variations of Hodge structure of very large dimension. It seems quite likely that these give variations of maximum possible dimension for the receiving period domain. A construction for even weight follows. Fix a period domain D with Hodge numbers $h^{r,s}$. Abbreviate these as $g^r = h^{r,s}$. Let B_r be the generalized unit ball associated to complex matrices of size $g^r g^{r-1}$. Viewing elements of B_r as weight one structures L with a distinguished symplectic complex structure J, form, for r even, the weight n structure $Q(L,r) = P(L,r) \otimes \mathrm{Triv}(n-r)$. Here $\mathrm{Triv}(m)$ is a Hodge structure of dimension one, even weight m, and type (m/2, m/2). Define a map

$$(4.6)\qquad v_M : Y = B_2 \times \dots \times B_n \longrightarrow D$$

by

$$(4.7)\qquad (H_2 , \dots , H_n) \longrightarrow \oplus\, Q(H_i, i),$$

where the indices in both the direct sum and the Cartesian product are even.

Remarks :

1. Although v_M is a degenerate variation of Hodge structure, it is maximal in the sense that if v' is another variation with image $(v') \subset$ image (v_M), then dim image (v') = dim image (v_M) . One may verify this by showing that the associated infinitesimal variation satisfies $T = Z(T)$, where $Z(T)$ is the centralizer of T (see [2]).

2. If the weight is $n = 2k$, then dimension of Y is roughly half that of the horizontal tangent bundle of D :

$$\dim T_h = \sum_{r \in I} g^r g^{r-1}$$

$$\dim T = \sum_{r \in I,\ r \text{ even}} g^r \approx \frac{1}{2} \dim T_h$$

where $I = [k+1, n]$.

3. The dimension of many (most ?) maximal variations is much smaller than $(1/2)\dim T_h$. According to the results of [5], the natural versal family of hypersurfaces of degree d and dimension n is maximal for $n \geq 2$, with the exception of cubics of dimensions 3, 4, and 5. If the canonical bundle of the hypersurface is non-nega tive, then the dimension of the variation (the dimension of the image of moduli in periods) is $g^n g^{n-1}$, the first term in the sum above.

4. The Hodge bundles F^p of v_M are flat for p odd, so that v_M is defined by Schubert conditions.

5. The Hodge structures in H in Y are characterized by the conditions (a) that there be a splitting $H \cong \oplus H_r$, (b) that H_r have type $\{ (r, n-r), (r-1, n-r+1), (n-r, r), (n-r+1, r-1) \}$ with r even, (c) that each term H_r admit an automorphism J with $J^2 = -1$ and (d) that this automorphism restrict to multiplication by $+i$ on the component of type $(r, n-r)$.

6. One can construct examples of maximal variations using $P(L, r)$ with r odd for which statements analogous to the above hold.

Bibliography

1. W.L. Baily and A. Borel : Compactification of arithmetic quotients of bounded symmetric domains, Ann. of Math. 84 (1966), 443-528.

2. J. A. Carlson, Bounds on the dimension of a variation of Hodge structure, Trans. AMS. 294 (1986) 45-64.

3. J. A. Carlson, M. L. Green, P.A. Griffiths, and J. Harris, Infinitesimal variations of Hodge structures I, Compositio Math. 50 (1983), pp. 109-205

4. J. A. Carlson, and P.A. Griffiths, Infinitesimal variations of Hodge structure and the global Torelli problem, Journées de Géometrie Algébrique d'Angers 1979, Sijthoff & Nordhoff, Alphen an den Rijn, the Netherlands (1980), pp 51-76.

5. J. A. Carlson and Ron Donagi, Hypersurface variations are maximal, preprint, July 1985.

6. P. A. Griffiths, Periods of rational integrals, I, II, Annals of Mathematics 90 (1969), 460-526 and 805-865.

7. P. A. Griffiths, Periods of integrals on algebraic manifolds, III, Publ. Math. I.H.E.S. 38 (1970), 125-180.

8. A. Grothendieck, Hodge's conjecture is false for trivial reasons, Topology 8 (1969) 299-303.

9. S. Helgason, Differential Geometry, Lie Groups, and Symmetric Spaces, Academic Press 1978, pp 634.

10. M. Kuga, Fiber varieties over a symmetric space whose fibers are abelian varieties, Proceedings of Symposia in Pure Mathematics, IX, American Math. Soc. 1966, 338-346.

11. D. Mumford, Families of Abelian Varieties, Proceedings of Symposia in Pure Mathematics, IX, American Math. Soc. 1966, 347-351.

12. D. Mumford, A Note on Shimura's Paper "Discontinuous Groups and Abelian Varieties", Math. Ann. 181 (1969), 345-351.

13. G. Shimura, Moduli of Abelian Varieties, Proceedings of Symposia in Pure Mathematics, IX, American Math. Soc. 1966, 312-332.

VARIATIONS OF POLARIZED HODGE STRUCTURE: ASYMPTOTICS AND MONODROMY

Eduardo Cattani[1], Aroldo Kaplan[1] and Wilfried Schmid[2]
University of Massachusetts
Amherst, MA 01003
Harvard University
Cambridge, MA 02138

§1. Introduction

In this article we describe the asymptotic properties of a variation of pure, polarized Hodge structure, degenerating along a divisor with normal crossings. Specifically, we shall discuss the structure of the local monodromy around the divisor and derive appropriate asymptotic formulae for the Hodge filtration and metric; it will be evident that these two aspects are very much interdependent. A principal application - the L_2 realization of Intersection cohomology with coefficients in such a variation - will be included in a companion paper in this same volume.

Although the results below are essentially contained in [CKS 1-2], here we have simplified and strengthened some of the constructions. In the present form they are better suited for other applications - variation of mixed Hodge structure, integral (p,p)-cycles - to be discussed elsewhere. We refer to [CKS 1-2] for full details and other consequences and to the article of Kashiwara [K-1] for another proof of the norm estimates.

Consider then a real, polarized variation of Hodge structure of weight k over a complex manifold X, $(\mathbf{V},\mathbf{V}_{\mathbf{R}},\mathbf{S},\mathbf{F})$. Here $\mathbf{V}$ is the underlying local system of finite dimensional complex vector spaces over X, $\mathbf{V}_{\mathbf{R}}$ the sheaf of real forms, $\mathbf{S}$ the polarization form and $\mathbf{F}$ the Hodge filtration on $\mathcal{O}_X \otimes \mathbf{V}$. By definition, $\mathbf{F}$ satisfies the transversality relation $\partial \mathbf{F}^p \subset \Omega^1_X \otimes \mathbf{F}^{p-1}$ and defines pure Hodge structures of weight k on the stalks of $\mathbf{V}$, polarized by the flat form $\mathbf{S}$.

X will be assumed to lie as the complement of a divisor with normal crossings in a complex manifold $\bar{X}$. Locally at the divisor, we may replace X by $(\Delta^*)^r \times \Delta^{n-r}$ and $\bar{X}$ by Δ^n, with Δ = unit disk, $\Delta^* = \Delta - (0)$, $n = \dim X$. For simplicity, we shall focus on the case when $r = n$ and the monodromy of $\mathbf{V} \longrightarrow (\Delta^*)^n$ is

(1) Partially supported by NSF Grant DMS-8501949
(2) Partially supported by NSF Grant DMS-8317436

unipotent (the first assumption can be made without essential loss of generality; for variations defined over $\mathbf{Z}$ - such as those arising in algebraic geometry - the same is true of the second).

The limiting mixed Hodge structure. According to the Nilpotent Orbit Theorem [S], the Hodge filtration extends to a filtration of the canonical extension of $\mathbf{V}$ over Δ^n. In other words, let $\mathcal{U}$ denote the upper half-plane, covering Δ^* via $z \longrightarrow e^{2\pi iz}$. Upon lifting to $\mathcal{U}$, the $\mathbf{F}^p$'s can be regarded as the pull-backs of universal bundles over

> D = classifying space of polarized Hodge structures on the typical fiber $V = \mathbf{V}_{\underline{s}_0}$, of the appropriate weight and Hodge numbers

via the "lifting of the period map" $\tilde{\phi}: \mathcal{U}^n \longrightarrow D$.

Let $\{\exp N_j\}$ be the monodromy transformations corresponding to counterclockwise simple circuits around the various punctures, acting on V. Then $\tilde{\phi}(z_1,\ldots,z_j+1,\ldots,z_n) = (\exp N_j)\tilde{\phi}(z_1,\ldots,z_j,\ldots,z_n)$ and therefore

$$(1.1) \qquad \tilde{\psi}(\underline{z}) = \exp(-\Sigma z_j N_j) \cdot \tilde{\phi}(z)$$

drops to a filtration $\psi(\underline{s}) = \tilde{\psi}(\frac{1}{2\pi i} \cdot \log \underline{s})$ depending holomorphically on $\underline{s} \in (\Delta^*)^n$. The Nilpotent Orbit Theorem asserts then that ψ extends holomorphically to all of Δ^n. Moreover, let

$$(1.2) \qquad F \overset{\text{def.}}{=} \psi(\underline{0}) = \lim_{\operatorname{Im} z_j \to \infty} \exp(-\Sigma z_j N_j) \cdot \tilde{\phi}(\underline{z})$$

be the *limiting Hodge filtration.* Then $\exp \Sigma z_j N_j \cdot F$ is pure and polarized for $\operatorname{Im} z_j \to \infty$ and the *nilpotent orbit* $\underline{z} \longmapsto \exp(\Sigma z_j N_j) \cdot F$ approximates $\tilde{\phi}$ in a sharp sense (cf. §3 below). It should be noted that $\psi(\underline{z})$, F, need not lie in D any longer, since $\exp(\Sigma z_j N_j)$ does not preserve $V_{\mathbf{R}}$. On the other hand, this transformation does preserve the form S, so that $\psi(\underline{z})$, F, can be regarded as lying in the "compact dual" $\hat{D}$ of D, consisting of all k-selfdual filtrations F:

$$(1.3) \qquad S(F^p, F^{k-p+1}) = 0$$

of the right weight and Hodge type.

This effectively reduces questions about the behavior of the variation near the divisor to properties of the limiting data V, $V_{\mathbf{R}}$, $\{N_j\}$, F. A priori

(1.4) The N_j's are commuting nilpotent endomorphisms of $(V, V_{\mathbf{R}}, S)$, F a k-selfdual filtration and $N_j F^p \subset F^{p-1}$ (transversality).

The one-variable version of the SL(2)-Orbit Theorem [S] coupled with the "independence of the weight filtration" of [CK-1] implies

(1.5) There exists a filtration W such that $(W[-k], F)$ is a mixed Hodge structure, polarized by the cone $C = \{\Sigma\lambda_j N_j | \lambda_j \in \mathbf{R}_+\}$ and the form S.

By definition, the polarization condition means: for all $N \in C$ and $\ell \in \mathbf{Z}$,

$$NW_\ell \subset W_{\ell-2}, \qquad N^\ell : Gr^W_\ell \xrightarrow{\sim} Gr^W_{-\ell}$$

(that is, W is the weight filtration of any N, $W(N)$); and the forms $S(\cdot, N^\ell \cdot)$ polarize the Hodge structures induced by F on the corresponding graded primitive parts

$$P_N Gr^W_\ell = \ker (N^{\ell+1} : Gr^W_\ell \longrightarrow Gr^W_{-\ell-2}).$$

Conversely, the conditions (1.4), (1.5) *characterize* the possible data $N_1, \ldots, N_n$; F, that can arise from a (real, unipotent) variation over $(\Delta^*)^n$, since they can be seen to imply that $\underline{z} \longrightarrow \exp(\Sigma z_j N_j) \cdot F$ is a nilpotent orbit (cf. (4.66) in [CKS-1]).

Fix now a subset $J \subset \{1, \ldots, n\}$ and an element $N \in \bar{C}$ such that

(1.6) $\mathrm{span}_{\mathbf{R}_+} \{C', N\} \subset C$, with $C' = \mathrm{span}_{\mathbf{R}} \{N_j | j \in J\}$;

such a N will be said to be *opposite* to the cone C'. Then $(z_j)_{j\in J} \longmapsto \exp(\sum_{j\in J} z_j N_j + zN) \cdot F$ is a nilpotent orbit for any $z \in \mathbf{C}$, $\mathrm{Im}\, z \gg 0$, with $\{N_j | j \in J\}$ as generators. Therefore

(1.7) $W' = W(C')$

is well defined and one has:

(1.8) Theorem. (i) $(W'[-k], \exp zN \cdot F)$ is a mixed Hodge structure for $\mathrm{Im}\, z \gg 0$, polarized by C'. (ii) W and F induce in $Gr^{W'}$ the limiting mixed Hodge structure of the variation (of pure structures) induced by $z \longmapsto \exp zN$; in particular, (iii) W is the weight filtration of N *relative* to W'.

Proof. $\exp zN \cdot F$ is the limiting mixed Hodge structure of the orbit generated by the N_j's, $j \in J$, so (i) is clear. (iii) is proved in [CK-1], while (ii) is a direct consequence of (i) and (iii).

(1.9) Remark. In (1.8) (iii), the pure structures in $Gr_\ell^{W'}$ are polarized by the forms $\sigma_\ell^{N'}$, $N' \in C'$, via the Lefschetz decomposition, $Gr_\ell^{W'} = \bigoplus_{r\geq 0} N'^r(P_{N'}Gr^{W'}_{\ell+2r})$: distinct summands are orthogonal and $\sigma_\ell^{N'}(u,v) = (-1)^r S(u,N'^\ell v)$, for u, v in the rth-summand. The statements of (1.8) describe then a variation of *mixed* Hodge structures, in which, except for the polarization statement of (i), no reference is made to W' being the weight filtration of a nilpotent endomorphism.

In the next section we construct distinguished **R**-gradings of the filtrations W', compatible with a given **R**-grading of W. They are given by a simple, explicit formula in terms of the data $N_1,\ldots,N_n;F$. In §3 we write down asymptotic expressions for $\tilde{\phi}$ in terms of those gradings. A subtler property of the monodromy is discussed in §4, reflecting the purity of the Hodge structures in the global Intersection cohomology groups.

§2. Gradings

Recall that for any mixed Hodge structure (W,F), the subspaces

$$(2.1)\qquad I^{p,q}(W,F) = F^p \cap W_{p+q} \cap (\bar{F}^q \cap W_{p+q} + \sum_{r\geq 1} \bar{F}^{q-r} \cap W_{p+q-r-1})$$

define the unique bigrading of (W,F):

$$W_\ell = \bigoplus_{p+q\leq \ell} I^{p,q}, \qquad F^p = \bigoplus_{r\geq p} I^{r,s}$$

satisfying

$$(2.2)\qquad I^{p,q} \equiv \overline{I^{q,p}} \qquad (\mathrm{mod} \bigoplus_{\substack{r<p\\ s<q}} I^{r,s}) \quad .$$

(cf. [CK-2], [D]). The subspaces

$$(2.3)\qquad I_\ell(W,F) = \bigoplus_{p+q=\ell} I^{p,q}(W,F)$$

grade W over **C**, compatibly with F and with any $(-r,-r)$ morphism of (W,F). Note that (W,F) splits over **R** iff the I_ℓ's are defined over **R**. In any case, the nilpotent subalgebra $\Lambda^{-1,-1}(W,F) = \{X \in \mathrm{End}\, V \mid XI^{p,q} \subset \bigoplus \{I^{r,s} \mid r<p, s<q\}\}$ is always defined over **R**.

For the next three Propositions, we shall assume that the "initial" mixed Hodge structure polarized by C,

$(W[-k],F)$, is split over **R**

(by the grading $U_\ell = I_\ell(W[-k],F)$), so as to facilitate the inductive argument. The reduction to this case will be discussed later.

Let now C', $W' = W(C')$, be associated to a given subset $J \subset \{1,\ldots,n\}$ and let $N \in \bar{C}$ be opposite to C', as in (1.6), so that

$$W, F, N \text{ satisfy (1.8) relative to } W'.$$

In particular, they determine **C**-gradings of $W'[-k]$, $\{I_\ell(W'[-k],\exp zN\cdot F)\}$.

(2.4) <u>Proposition</u>. The grading of W' given by

$$U'_\ell = \exp(-zN) \cdot I_{k+\ell}(W'[-k],\exp zN\cdot F)$$

for $\operatorname{Im} z >> 0$, is independent of z, defined over **R** and compatible with the filtration F, the **R**-grading U of W and the polarizing form S.

The next two Propositions show that this construction can be carried out inductively for any descending chain of subsets J, yielding mutually compatible gradings of the associated weight filtration. Since N preserves W', we can write

(2.5) $\qquad \exp zN = \exp \Gamma(z) \exp z\hat{N}$

where $\hat{N}$ is the homogeneous component of weight 0 relative to U' and $\Gamma(z) \in W'_{-1}\mathfrak{g}$ ($\mathfrak{g}$ = Lie algebra of infinitesimal isometries of (V,S)). Keeping the hypothesis and notation of (2.4), (2.5), we have

(2.6) <u>Proposition</u>. (i) The **R**-split mixed Hodge structure $(W[-k],F)$ is also polarized by the $\mathbf{R}_+$-span of C' and $\hat{N}$.

(ii) For $\operatorname{Im} z >> 0$, $(W'[-k],\exp z\hat{N}\cdot F)$ is a mixed Hodge structure, polarzed by C' and split over **R** by U'.

(iii) $\Gamma(z) \in \Lambda^{-1,-1}(W'[-k],\exp zN\cdot F)$ and it commutes with all morphisms of this mixed Hodge structure.

Let now $K \subset J \subset \{1,\ldots,n\}$, C'' the $\mathbf{R}_+$-span of $\{N_k | k \in K\}$, $W'' = W(C'')$ and let $N' \in \bar{C}'$ be opposite to C'. With $\hat{N}$ as in (2.5), $N' + \hat{N}$ is opposite to C'' in their common $\mathbf{R}_+$-span.

(2.7) <u>Proposition</u>. The two sets of data $(W,F,N'+\hat{N})$ and $(W',\exp i\hat{N}\cdot F,N')$ determine - via (2.4) - the same **R**-grading U'' of W'', which is then compatible with both U and U'.

Propositions (2.4), (2.6), (2.7) contain the algebraic statements of the several-variables version of the SL(2)-orbit Theorem of [CKS-I], somewhat reshaped. We now give a sketch of their

<u>Proof</u>: The argument is based on the analysis of the nilpotent orbit

$$t \longmapsto \exp itN'\cdot F', \qquad F' = \exp iN\cdot F$$

determined by any fixed $N' \in C'$. The one-variable version of the SL(2) orbit theorem ([S]) already gives a representation

$$\exp itN'\cdot F' = g(t)\exp itN'\cdot\hat{F}'$$

where: $t \longmapsto \exp itN'\cdot\hat{F}'$ is a nilpotent orbit whose limiting mixed Hodge structure $(W'[-k],\hat{F}')$ splits over $\mathbf{R}$; $g(t)$ is a $G_{\mathbf{R}}$-valued function having an expansion $g(t) = 1 + \frac{1}{t}g_1 + \frac{1}{t^2}g_2 + \ldots$ for $t \gg 0$; and $g_\ell \in W'_{\ell+1}\,\mathrm{End}\,V$ (in the notation of [S], our N', F', $\hat{F}'$, are, respectively, N, F, $\exp(-iN)\cdot 0$). The last property implies that the limit $\lim_{t\to\infty}\exp(-itN')g(t)\exp(itN') \overset{\text{def.}}{=} \gamma$ exists and lies in $\exp W_{-1}\mathfrak{g}$. Clearly,

(2.9) $\gamma\cdot\hat{F}' = F'$ $(= \exp iN\cdot F)$

and an additional argument (cf. [CK-2] or [CKS-1]) shows that

(2.10) $\gamma \in \exp\Lambda^{-1,-1}(W',F')$ and commutes with $(-r,-r)$-morphisms.

We now use the special nature of the nilpotent orbit (2.8) as arising from a two-variable orbit generated by N and N'.

For any polarizable mixed Hodge structure of weight k, $(\tilde{W}[-k],\tilde{F})$, we let $T(\tilde{W},\tilde{F})$ denote the semisimple element of $\mathrm{Env}\,V$ such that

(2.11) $T(\tilde{W},\tilde{F}) = (\ell - k)1$ on $I_\ell(\tilde{W}[-k],\tilde{F})$.

Note that $(\tilde{W}[-k],\tilde{F})$ splits/ $\mathbf{R}$ iff $T(\tilde{W},\tilde{F})$ is real. This is true then of the elements

(2.12) $Y = T(W,F)$ and $Y' = T(W',\hat{F}')$.

One now shows that Y commutes with both Y' and $g(t)$ and

(2.13) $tN' + N = \mathrm{Ad}(g(t))\,(tN' + N_0)$,

where N_0 is the 0-weight component of N relative to the grading by eigenspaces of Y'; this is Lemma (4.31) in [CKS-1]. The last identify yields $\exp iN = (\exp(-itN')g(t)\exp itN')\exp itN_0 g(t)^{-1}$ and, letting $t\to\infty$,

(2.14) $\exp iN = \gamma\exp iN_0$.

The uniqueness of the bigradings $\{I^{p,q}\}$ easily implies

(2.15) $\mathrm{Ad}(p)T(\tilde{W},\tilde{F}) = T(\tilde{W},p\cdot\tilde{F})$, for

$$p \in \exp\tilde{W}_0\,\mathrm{End}\,V_{\mathbf{R}} \quad\text{or}\quad p \in \exp\Lambda^{-1,-1}(\tilde{W}[-k],\tilde{F}) .$$

In particular, $\mathrm{Ad}(\gamma)Y' = T(W',\gamma\cdot\hat{F}') = T(W',\exp iN\cdot F)$. Since $\exp iN_0$ centralizes Y', (2.14) gives $\mathrm{Ad}(\gamma)Y' = \mathrm{Ad}(\exp iN)Y'$ and, therefore, $Y' = \mathrm{Ad}(\exp -iN)$ $T(W',\exp iN\cdot F)$. This shows (2.4) for $z = i$, since the right-hand side has U'_ℓ as eigenspaces. The statement for arbitrary $z = x + iy$, $y > 0$, follows from (2.15), the formula $\exp zN = \exp xN\exp(-\frac{1}{2}\log y\,Y)\exp iN(\exp\frac{1}{2}\log y\,Y)$ and the fact that $N, Y \in W_0'$ End $V_{\mathbb{R}}$ and $YF^p \subset F^p$.

For (2.6), note that $\hat{N}$ and $\exp \Gamma(i)$ agree with the elements N_0, Y, above. Hence (i) follows from (2.5), (2.10), (2.14). Since N commutes with every element of C' and these are homogeneous relative to U', so does $\hat{N}$. By definition, U', $\hat{N}$, do not depend on the particular element N'. From (2.13) one deduces: $W(N' + \hat{N}) = W(N' + N) = W$ for $N' \in C'$; and that $\exp(z'N' + z\hat{N}) \cdot F$ is pure and polarized for $\operatorname{Im} z$, $\operatorname{Im} z' >> 0$. This gives (2.6)(ii). Finally, (iii) follows from (2.4), (2.10) and (2.15).

To prove (2.7), we shall deal again with the semisimple elements T associated to the gradings in question. Then $Y = T(W,F)$ grades W, $Y' = \operatorname{Ad} \exp(-iN)$ $T(W', \exp iN \cdot F)$ grades W' and the elements $Y'' = \operatorname{Ad} \exp(-i(\hat{N} + N'))\, T(W'', \exp i(\hat{N} + N') \cdot F)$ and $\tilde{Y}'' = \operatorname{Ad}(\exp iN')\, T(W'', \exp iN' \cdot (\exp i\hat{N} \cdot F))$ grade W'' and correspond in that order to two sets of data in the statement of (2.7). Clearly, $\tilde{Y}'' = \operatorname{Ad}(\exp i\hat{N})Y''$. Due to the relations $[Y,N] = -2N$, $[Y',Y] = 0$ and the definition of $\hat{N}$, one has: $[Y,\hat{N}] = -2\hat{N}$. Also, Y'' commutes with Y. Therefore

$$0 = \operatorname{Ad}(\exp i\hat{N})[Y'',Y] = [\tilde{Y}'',Y + 2i\hat{N}] = [\tilde{Y}'',Y] + 2i[\tilde{Y}'',\hat{N}]$$

implying $[\tilde{Y}'',\hat{N}] = 0$. But then $Y'' = \operatorname{Ad}(\exp -i\hat{N})\tilde{Y}'' = \tilde{Y}''$. Also, $\hat{N}$ is already U''-homogeneous of weight zero. This ends the proof of (2.5)-(2.7).

In order to complete the inductive construction of $\mathbb{R}$-gradings it is necessary to consider the case when the initial $(W[-k],F)$ does not split over $\mathbb{R}$. There are various ways of doing this. In [CKS-1], the correspondence $(W'[-k], \exp iN \cdot F) \to (W'[-k], \exp i\hat{N} \cdot F)$ arises naturally as a special case of a general one

$$\{\text{mixed Hodge structures}\} \to \{\mathbb{R}\text{-split mixed Hodge structures}\}$$

which respects morphisms; it has the relative disadvantage that a simple formula such as that in (2.4) is not available in general. On the other hand, the following simple procedure (Deligne) follows directly from the properties of the bigradings $I^{p,q}$. Given any mixed Hodge structure (W,F) there exist a unique $\delta \in \Lambda^{-1,-1}(W,F)$ such that $\exp(-2i\delta) \cdot F^p = \bigoplus_{r \geq p} I^{s,r}$ (recall that $F^p = \bigoplus_{r \geq p} I^{r,s}$). This element must be then real, commute with all morphisms of (W,F) and, with

(2.16) $\quad \hat{F} = \exp(-i\delta) \cdot F,$

the mixed Hodge structure $(W,\hat{F})$ splits over $\mathbb{R}$ (cf. [D], [CK-2]).

Consider again the limiting data of a variation $(N_1,\ldots,N_n)$, F, with the given ordering of the monodromy logarithms. Inductively, the construction above yields mutually compatible $\mathbb{R}$-gradings U^j of the weight filtrations $W^j = W(N_1,\ldots,N_j)$, $1 \leq j \leq n$,

$$W_\ell^j = \bigoplus_{m\leq\ell} U_m^j .$$

Explicitly, let $\hat{F} = \exp(-i\delta)\cdot F$ be as in (2.16) and define

(2.17) $U_\ell^n = I_{k+\ell}(W^n[-k],\hat{F})$,

U^j = grading of W^j determined as in (2.4) by the objects W^{j+1}, $\exp i\hat{N}_{(j+2)}\cdot\hat{F}$, N_{j+1}, $(j < n)$,

$\hat{N}_{(n+1)} = 0$,

$\hat{N}_{(j)}$ = homogeneous component of weight 0 of $N_j+\dots+N_k$ relative to U^{j-1}.

Then,

(2.18) $\hat{F}_0 \overset{\text{def.}}{=} \exp i\hat{N}_{(n)}\cdot F$ lies in D, and the multigrading $V = \oplus\, U_{\ell_1,\dots,\ell_n}$, $U_{\ell_1,\dots,\ell_n} = \bigcap_j U^j_{\ell_j}$, is orthogonal with respect to the Hodge form at $\hat{F}_0$, $S(C_{\hat{F}_0}\,\cdot,\cdot)$.

Define elements $\Gamma_0(=0), \Gamma_1,\dots,\Gamma_n$ by: $\Gamma_n = i\delta$ and $\exp N_{j+1} = \exp\Gamma_j\cdot\exp\hat{N}_{j+1}$, according to U^j, as in (2.5). The following Lemma collects the properties of the objects U^j, Γ_j, N_j, $\hat{N}_j$, needed in the next section; they are immediate consequences of (2.4-7).

(2.19) Lemma. (i) N_j, $\hat{N}_j$ are homogeneous of weight -2 relative to $U^j,\dots,U^n$. (ii) $\hat{N}_j$ is homogeneous of weight 0 relative to $U^1,\dots,U^{j-1}$. (iii) Γ_j commutes with $N_1,\dots,N_j$ and with $\hat{N}_1,\dots,\hat{N}_j$. (iv) Γ_j is of weight 0 relative to $W^1,\dots,W^{j-1}$. (v) Γ_j is of weight -2 relative to W^j.

§3. Asymptotic behavior of the Hodge filtration and norm estimates

We first derive an expression for the Hodge filtration along a nilpotent orbit, $\exp(\Sigma z_j N_j)\cdot F$, adapted to the *ordered* set $(N_1,\dots,N_n)$. Recall (2.17), (2.18), in particular the **R**-multigrading $V = \oplus\, U_{\ell_1,\dots,\ell_n}$ and the pure structure $\hat{F}_0 \in D$. Define a $G_{\mathbf{R}}$-valued function $e(\underline{y})$ by

(3.1) $e(\underline{y})$ acts with eigenvalue $\left(\frac{y_1}{y_2}\right)^{\ell_1/2}\cdots\left(\frac{y_{n-1}}{y_n}\right)^{\ell_{n-1}/2} y_n^{\ell_n/2}$ on $U_{\ell_1,\dots,\ell_n}$,

and a $G_{\mathbf{C}}$-valued function $p(\underline{y}) = e(\underline{y})\exp(i\Sigma y_j N_j)\exp(i\delta)e(\underline{y})^{-1}\exp(-i\hat{N}_{(n)})$. A priori, p is a Laurent polynomial in the variables $y_1^{1/2},\dots,y_n^{1/2}$.

(3.3) Theorem. (i) $\exp i(\Sigma y_j N_j)\cdot F = e(\underline{y})p(\underline{y})\cdot \hat{F}_0$, with $p(\underline{y}) \to 1$ as $\frac{y_1}{y_2},\ldots,\frac{y_{n-1}}{y_n}$, $y_n \to \infty$; more specifically,

(ii) $p(\underline{y}) = \prod_{j=1}^{n} \exp\left(\left(\frac{y_j}{y_{j+1}}\right)^{-1} P_j(\underline{y})\right)$, where $P_j(\underline{y})$ is a polynomial function in the ratios $\left(\frac{y_1}{y_2}\right)^{-\frac{1}{2}},\ldots,\left(\frac{y_j}{y_{j+1}}\right)^{-\frac{1}{2}}$ (*), with values in $(W^j_{-2}\mathfrak{g}) \cap \bigcap_{r=1}^{n} \ker \operatorname{ad} N_r$.

Proof. Since $\hat{F}$ is compatible with $U_{\ell_1,\ldots,\ell_n}$, $e(\underline{y})$ stabilizes it, so that (i) follows from (2.18), (2.16). Let $e_j(t)$ operate on U^j_ℓ as $t^{\ell/2}1$, so that $\operatorname{Ad}(e_j(t))N_r = t^{-\frac{1}{2}}N_r$ for $j \geq r$, and set $P_j(\underline{y}) = \frac{y_j}{y_{j+1}} \operatorname{Ad}\left(e_1\left(\frac{y_1}{y_2}\right)\cdots e_j\left(\frac{y_j}{y_{j+1}}\right)\right)D_j$.

To verify (ii), one decomposes $\exp i\Sigma y_j N_j = \exp iy_n\left(\sum_{j=1}^{n-1} \frac{y_j}{y_n} N_j + N_n\right) =$ $e_n(y_n)^{-1} \exp i\left(\sum_{j=1}^{n-1} \frac{y_j}{y_n} N_j\right)\exp iN_n\, e_n(y_n)$ and $\exp iN_n = \exp \Gamma_{n-1} \exp i\hat{N}_n$, obtaining $\exp i(\Sigma y_j N_j) \exp \Gamma_n = e_n(y_n)^{-1} \exp i\left(\sum_{j=1}^{n-1} \frac{y_j}{y_n} N_j\right)\exp \Gamma_{n-1} \exp(\operatorname{Ad}(e_n(y_n)))\Gamma_n \exp \hat{N}_n e_n(y_n)$.
Inductively, and using (2.19) one arrives to

$$(3.4) \qquad \exp i(\Sigma y_j N_j)\exp \Gamma_n = e(\underline{y}) \prod_{j=1}^{n} \exp\left(\frac{y_j}{y_{j+1}}\right)^{-1} P_j(\underline{y}))\exp i\hat{N}_{(n)} e(\underline{y}),$$

which proves the identity in (ii) ($\Gamma_n = i\delta$ by definition). The remaining property of the P_j's follows from (2.19)(iv),(v).

Consider now the map $\tilde{\phi}:\mathcal{U}^n \longrightarrow D$, whose value at $\underline{z} = \frac{1}{2\pi i} \log \underline{s}$, $(0) \subset \tilde{\phi}^k(\underline{z}) \subset \ldots \subset \tilde{\phi}^o(\underline{z}) = V$ $(= \mathbf{V}_{s_0})$ represents the Hodge filtration on the fiber at $\underline{s}$. One has the following asymptotic representation for $\tilde{\phi}$, good on regions adapted to the given ordering in $(N_1,\ldots,N_n)$.

(*) We set $y_{n+1} \equiv 1$, so these statements hold for $j = n$ also.

(3.5) Theorem. (i) $\tilde{\phi}(\underline{x} + i\underline{y}) = \exp(\Sigma x_j N_j) e(\underline{y})^{-1} p(\underline{y}) q(\underline{x},\underline{y}) \cdot \hat{F}_0$, where $\hat{F}_0 \in D$ and $p(\underline{y}) \in G_{\mathbb{C}}$ are as in (3.3), $q(\underline{x},\underline{y}) \in G_{\mathbb{C}}$ is defined for $y_j \gg 0$ and $q \to 1$ as $\frac{y_1}{y_2}, \ldots, y_n \to \infty$. More precisely, $q(\underline{x},\underline{y}) = \exp Q(\underline{x},\underline{y})$, with

(ii) $Q(\underline{x},\underline{y}) \in \mathfrak{g}$ is nilpotent and real analytic in $e^{2\pi i \underline{x}}$, $\underline{y}$,

(iii) for any $\varepsilon > 0$ there are positive constants C, c, K such that $\|Q(\underline{x},\underline{y})\| < Ce^{-cy_n}$ for $y_1 > \varepsilon y_2, \ldots, y_{n-1} > \varepsilon y_n$, $y_n > K$ ($\| \ \|$ a fixed linear norm on End V). A similar estimate applies to the derivatives $\prod_j \left(y_j \frac{\partial}{\partial y_j}\right)^{m_j} Q$, $\prod_j \left(y_j \frac{\partial}{\partial_j}\right)^{m_j} Q$.

(3.6) Corollary. For any $\varepsilon > 0$, the filtrations $e(\underline{y})\exp(-\Sigma x_j N_j)\tilde{\phi}(\underline{x} + i\underline{y})$ lie in a compact subset of D for $y_1 > \varepsilon y_2, \ldots, y_{n-1} > \varepsilon y_n, y_n > \varepsilon$. The same is true of $e(\underline{y})\tilde{\phi}(\underline{x} + i\underline{y})$ if $\underline{x}$ remains bounded.

The subspaces $U_{\ell_1,\ldots,\ell_n}$ and the form $S(C_{\hat{F}_0} \cdot, \bar{\cdot})$ determine, locally, a flat $\mathbb{R}$-multigrading $\mathbf{V} = \oplus \mathbf{U}_{\ell_1,\ldots,\ell_n}$ and a flat, compatible hermitian norm $| \ |_0$ on $\mathbf{V} \to (\Delta^*)^n$. In particular, $\mathbf{W}_\ell(N_1,\ldots,N_r) = \bigoplus_{\ell_r \le \ell} \mathbf{U}_{\ell_1,\ldots,\ell_n}$.

(3.7) Corollary. On any region of the form

$$\{\underline{s} \in (\Delta^*)^n \,\big|\, |\underline{s}| < r, \ \frac{\log|s_j|}{\log|s_{j+1}|} \le \varepsilon, \ 1 \le j \le n-1\}, \quad r < 1, \quad \varepsilon > 0,$$

the Hodge metric $\| \ \|^2_{\underline{s}} = S(C_{\phi(\underline{s})}\cdot, \bar{\cdot})$, up to a quasi-isometry, satisfies: the decomposition $\mathbf{V} = \oplus \mathbf{U}_{\ell_1,\ldots,\ell_n}$ is orthogonal and

$$\|v\|^2 = \left(\frac{\log|s_1|}{\log|s_1|}\right)^{\ell_1} \cdots \left(\frac{\log|s_{n-1}|}{\log|s_n|}\right)^{\ell_{n-1}} (-\log|s_n|)^{\ell_n} |v|_0^2$$

for any locally flat section $v \in \mathbf{U}_{\ell_1,\ldots,\ell_n}$.

(3.7) Remarks. (3.3) and (3.5) correspond to (5.10) in [CKS-1]. The present version incorporates some improvements. The regions of validity are strictly larger, exponential and polynomial behaviors have been clearly separated by the functions p and q and the somewhat unpleasant rings appearing in [ibid] have been eliminated. Although, for the norm estimates, (3.5) suffices, the additional information provided by (3.4) is needed in other applications. For example, the higher order estimates contained in (i), (v) are used in proving that the Chern classes of the canonical extension $\bar{\mathbf{V}} \to \bar{X}$ are computed by the Chern forms of the Hodge metric of $\mathbf{V} \to X$. ([CKS-1]), [Ko]); and the distinction between polynomial and exponential behaviors enters in the proofs of finiteness and invariance of integral elements of type (p,p) near the divisor.

Proof of (3.5)-(3.7). Write $\tilde{\phi}(\underline{z}) = \exp(\Sigma z_j N_j)\cdot\psi(e^{2\pi i \underline{z}})$ with $\psi:\Delta^n \to G_{\mathbb{C}}$ holomorphic and $\psi(\underline{0}) = F$, as in (1.1), (1.2), and let $\hat{F} = \exp(-i\delta)\cdot F \in \check{D}$ be as in (2.17). $G_{\mathbb{C}}$ acts holomorphically and transitively on $\check{D}$, the stabilizer of $\hat{F}$ in $\mathfrak{g}$ is the subalgebra $\mathfrak{g}_{\hat{F}} = \bigoplus_{p\geq 0} I^{p,q}\mathfrak{g}$ (with $I^{p,q}$ as in (2.1)) and the nilpotent subalgebra $\mathfrak{v} = \bigoplus_{p\leq 0} I^{p,q}\mathfrak{g}$ is a linear complement of $\mathfrak{g}_{\hat{F}}$. One can then write $\psi(\underline{s}) = \exp i\delta \exp X(\underline{s})\cdot\hat{F}$ with $X(\underline{s}) \in \mathfrak{v}$ holomorphic for $\underline{s} \sim 0$ and $X(\underline{0}) = 0$. set

$$(3.9) \qquad Q(\underline{x},\underline{y}) = \mathrm{Ad}(\exp(i\hat{N}_{(n)})e(\underline{y}))X(e^{2\pi i \underline{z}}) .$$

With this Q, (i) follows from the definition of p and the relations $\hat{F}_0 = \exp i\hat{N}_{(n)}\cdot\hat{F}$, $e(y)\cdot\hat{F} = \hat{F}$; and (ii) from the nilpotency of $\mathfrak{v}$. To show (iii) we use the estimate of the Nilpotent Orbit Theorem (cf. (1.15) in [CKS-1], which contains the stronger estimate needed here): $d(\tilde{\phi}(\underline{z}), \exp \Sigma z_j N_j\cdot F) < \Sigma e^{-c_j y_j}$ for $y_j >> 0$, d a $G_{\mathbb{R}}$-invariant distance on D, c_j positive constants. Because $\exp \Sigma x_j N_j$, $e(\underline{y})$ lie in $G_{\mathbb{R}}$, (3.3)(i), (3.5)(i) yield

$$(3.10) \qquad d(p(\underline{y})\cdot\hat{F}_0,\ p(\underline{y})q(\underline{x},\underline{y})\cdot\hat{F}_0) < \Sigma e^{-c_j y_j}, \quad \text{for } y_j >> 0 .$$

As $\frac{y_1}{y_2},\dots,y_n \to \infty$, $p \to 1$, therefore $q\cdot\hat{F}_0 \to \hat{F}_0$ and $\exp(\mathrm{Ad}(e)X)\cdot\hat{F} \to \hat{F}$; since $\mathfrak{v}$ is e-invariant, $\mathrm{Ad}(e)X \in \mathfrak{v}$ and therefore $\mathrm{Ad}(e)X \to 0$. Let $X(\underline{s}) = \Sigma X_{\underline{m}}\underline{s}^{\underline{m}}$ be the expansion of X for $\underline{s} \sim 0$. Because of the definition of $e(\underline{y})$ and the fact that X was chosen independently of the ordering of the variables, this last assertion implies

$$(3.11) \qquad X_{\underline{m}} \in \cap\{W_0(N_{j_1},\dots,N_{j_r})\mathfrak{g} \mid m_{j_1} = \dots = m_{j_r} = 0\} . \quad (*)$$

It is now easy to see that $\mathrm{Ad}(e)X$ and, therefore, Q must satisfy the estimates (iii). For (3.6), note that $e\cdot\exp(-\Sigma x_j N_j)\tilde{\phi} = pq\cdot\hat{F}_0$ remains in a neighborhood of $\hat{F}_0 \in \check{D}$ at least for ε very large, since $p,q \to 1$. To cover the region for arbitrary ε one proceeds inductively, usinq the bounded ratios as parameters;

(*) Using transversality one can actually show: $[X_{\underline{m}}, N_j] = 0$ if $m_j = 0$.

we omit the details. The statement for $e\cdot\tilde{\phi} = \exp(\mathrm{Ad}(e)\Sigma x_jN_j)\cdot e\cdot\exp(-\Sigma x_jN_j)\tilde{\phi}$ follows from the first together with $\mathrm{Ad}(e)\Sigma x_jN_j = y_n^{-1}\,\mathrm{Ad}\left(e_1\left(\frac{y_1}{y_2}\right)\cdots e_{n-1}\left(\frac{y_{n-1}}{y_n}\right)\right)(\Sigma x_jN_j) = y_n^{n-1}\cdot(\text{polynomial in } \frac{y_1}{y_2},\ldots,\frac{y_{n-1}}{y_n}, x)$. Finally, (3.7) is a direct consequence of (3.5) and (2.18): since $e(\underline{y}) \in G_{\mathbf{R}}$, one has $C_{\tilde{\phi}} = e^{-1}C_{e\cdot\tilde{\phi}}\cdot e$, and therefore $h_{\underline{z}}(u,v) = S(C_{e\cdot\tilde{\phi}}e\cdot u, e\cdot\bar{v}) \sim S(C_{\hat{F}_0}\, e\cdot u, e\cdot\bar{v})$ on the required region.

§4. Further properties of the local monodromy: purity

In this section we will deduce some subtler properties of the local monodromy of a polarized variation of Hodge structure. These are encoded in a vanishing theorem for the cohomology of a complex which arises as the Intersection cohomology of Δ^n with values in the polarized variation of Hodge structure; they play a crucial role in the L_2 realization of this cohomology, as explained in our companion article.

As we have seen in §2, the local monodromy data of a polarized variation of Hodge structure of weight k, gives a commuting set $\{N_1,\ldots,N_n\}$ of nilpotent infinitesimal automorphisms of $(V_{\mathbf{R}}, S)$, and a filtration $F \in \hat{D}$, such that $(W[-k],F)$ is a mixed Hodge structure split over $\mathbf{R}$, and polarized by every $N \in C$, the positive linear span of $N_1,\ldots,N_n$. As before, W denotes the common weight filtration of the elements in C.

The commutativity of the N_j's implies that if $T_1, T_2 \in \mathrm{span}_{\mathbf{R}}\, N_1,\ldots,N_n$ then $T_1(T_2V) \subset T_2V$ and hence T_1 defines a nilpotent element $\tilde{T}_1 \in \mathrm{End}(T_2V)$. Although one may see from examples that, generally, the weight filtrations of T_1 and $\tilde{T}_1$ are not well related, the polarization conditions imposed on the cone C yield

(4.1) <u>Theorem</u>. Given $N \in C$ and $T \in \bar{C}$, the closure of C, let $\tilde{N} \in \mathrm{End}(TV)$ denote the restriction of N to TV. Then

(i) The weight filtration $W(\tilde{N})$ is given by $W_\ell(\tilde{N}) = TW_{\ell+1}$. In particular, this weight filtration is independent of $N \in C$; we set $\tilde{W}_\ell = TW_{\ell+1}$.

(ii) The filtrations $\tilde{W}[1-k]$ and $\tilde{F}$, $\tilde{F}^{p+1} = TF^p$, define a mixed Hodge structure, split over $\mathbf{R}$ and polarized by $\tilde{N}$ and the non-degenerate bilinear form $\tilde{S}$ defined on TV by

$$\tilde{S}(Tu, Tv) = S(u, Tv).$$

We give a sketch of the proof of (4.1). By restricting attention to the subcone spanned by N and T, we may assume without loss of generality that $n = 2$ and $T = N_1$. Since N_1 is a $(-1,-1)$ morphism of the mixed Hodge structure $(W[-k],F)$, a direct application of the strictness property of morphisms yields that $(\tilde{W}[1-k],\tilde{F})$ is an $\mathbf{R}$-split mixed Hodge structure on N_1V. Moreover, once (i) is established, it suffices to prove the polarization statement in (ii) for <u>some</u> $N \in C$, since the constancy of the weight filtration would them imply the conclusion for every such N.

A rigidity argument for mixed Hodge structures (cf. [CK-1], (2.17)) shows that if $W(\tilde{N})[1-k]$ also defines a mixed Hodge structure with $\tilde{F}$, then necessarily $W(\tilde{N}) = \tilde{W}$. Thus, it suffices to show:

(4.2) For every $N \in C$, $(W(\tilde{N})[1-k],\tilde{F})$ is a mixed Hodge structure.

For a "split" nilpotent orbit, the asymptotic expansion (3.3) takes on a particularly simple form; thus, we may write for $y_2 > 0$, $t = y_1/y_2 > 0$,

$$\exp(iy_2(tN_1 + N_2))\cdot F = e(\underline{y})^{-1}\cdot p(y)\cdot \hat{F}_0$$

$$= \exp((-1/2)\log(y_2)Y_2)\cdot\exp((-1/2)\log(t)Y_1)\cdot p_t\cdot\exp(iN_1)\cdot F_1$$

where $F_1 \in \hat{D}$, $\exp(iN_1)F_1 \in D$, $(W(N_1)[-k],F_1)$ is a mixed Hodge structure split over $\mathbf{R}$, Y_2 and Y_1 are the canonical commuting splittings of $(W[-k],F)$ and $(W(N_1)[-k],F_1)$, respectively, and $p_t = \exp(\mathrm{Ad}(\exp(1/2)\log(t)Y_1)\cdot\Gamma_1) \in W_{-2}(N_1)\mathfrak{g} \cap \ker N_1$ satisfies

$$\lim_{t\to\infty} p_t = 1 \quad .$$

We can similarly write

$$\exp(iy_2\tilde{N}(t))\cdot\tilde{F} = \exp((-1/2)\log(y_2)\tilde{Y}_2)\cdot\exp((-1/2)\log(t)\tilde{Y}_1)\cdot\tilde{p}_t\cdot\exp(i\tilde{N}_1)\cdot\tilde{F}_1$$

where $\tilde{N}(t)$ (resp. $\tilde{p}_t$) denote the actions of $tN_1 + N_2$ (resp. p_t) on N_1V and $\tilde{Y}_2$ (resp. $\tilde{Y}_1$) is the canonical splitting of the mixed Hodge structure $(\tilde{W}[1-k],\tilde{F})$ (resp. $(W(\tilde{N}_1)[1-k],\tilde{F}_1)$). Hence for t sufficiently large and $\mathrm{Im}(z) > 0$, $\exp(z\tilde{N}(t))\tilde{F}$ is a Hodge structure of weight $k-1$ on N_1V and thus the map

$$z \longrightarrow \exp(z\tilde{N}(t))\tilde{F}$$

is a nilpotent orbit. This implies that $(W(\tilde{N}(t))[1-k],\tilde{F})$ is a mixed Hodge structure polarized by $\tilde{N}(t)$, which establishes (4.2) - and a fortiori the polarization statement in (4.1;ii) - for any element of the form $N(t) = tN_1 + N_2$ with t sufficiently large.

In order to show that (4.2) holds for every $N \in C(N_1,N_2))$ it is now enough to prove that if (4.1) holds for every $N \in C(N_1,N(t_0))$, $t_0 > 0$, then

(4.3) $\tilde{W} = W(\tilde{N}(t_0))$.

These assumptions imply that $(\tilde{N}_1,\tilde{N}(t_0),\tilde{F})$ defines a nilpotent orbit on N_1V and consequently we may apply (1.8) to the action of $\tilde{N}_1$ on $Gr_\ell(W(\tilde{N}(t_0)))$. This space carries a natural mixed Hodge structure whose weight filtration is given by a suitable shifting of the projection of $\tilde{W}[\ell]$ to $Gr_\ell(W(\tilde{N}(t_0)))$. The failure of (4.3) is then reflected on the non-triviality of this projection. A comparison of the polarization properties of this mixed Hodge structure and those of $(W(N(t_0))[-k],F)$ leads to a contradiction. This is the content of Lemma (2.30) in [CKS-2].

Given a set of indices $J = (j_1,\ldots,j_p)$; $1 \leq j_1 < \ldots < j_p \leq n$ we set

$$B_J = N_{j_1}N_{j_2}\ldots N_{j_p}V; \qquad B^p = \bigoplus_{|J|=p} B_J$$

and define $d: B^p \longrightarrow B^{p+1}$ by:

$$(du)_J = \sum_{k=1}^{p+1} (-1)^{k-1} N_{j_k} u_{J-\{j_k\}}$$

where $u \in B^p$ and $(J = j_1,\ldots,j_{p+1})$. We bi-filter the complex (B^*,d) by setting

$$\tilde{W}_r(B_J) = N_{j_1}\ldots N_{j_p} W_{r+p}, \qquad \tilde{W}_r(B^p) = \bigoplus_{|J|=p} \tilde{W}_r(B_J)$$

$$\tilde{F}^r(B_J) = N_{j_1}\ldots N_{j_p} F^{r+p}, \qquad \tilde{F}^r(B^p) = \bigoplus_{|J|=p} \tilde{F}^r(B_J).$$

We observe also that B^* carries a natural non-degenerate bilinear form $\tilde{S}_p$ defined on B^p by

$$\tilde{S}_p(N_{j_1}\ldots N_{j_p}u, N_{j_1}\ldots N_{j_p}v) = S(u, N_{j_1}\ldots N_{j_p}v).$$

Successive applications of (4.1) give:

(4.4) $(\tilde{W}(B^p)[p-k], \tilde{F}(B^p))$ is a mixed Hodge structure on B^p, split over $\mathbf{R}$ and polarized by the action of every $N \in C$. The differential $d: B^p \longrightarrow B^{p+1}$ is a $(-1,-1)$-morphism of mixed Hodge structures.

The following vanishing theorem for the cohomology of the complex (B^*,d) was conjectured by Deligne. The special case $n = 2$, which already implies Theorem (4.1) was announced in [CK-3]. The statement (1.10) in [CKS-2] is somewhat stronger than the one below.

(4.5) Purity Theorem: Relative to the natural filtration on $H^*(B^*)$ induced by $\tilde{W}(B^*)$: $H^*(B^*) \subset \tilde{W}_0(H^*(B^*))$.

We conclude this section with a brief sketch of the proof of (4.5). Consider the natural real grading $\tilde{U}(B^p)$ of $\tilde{W}(B^p)$ associated to the $\mathbf{R}$-split mixed Hodge structure $(\tilde{W}(B^p)[p-k],\tilde{F}(B^p))$, i.e.

$$(4.6)\qquad \tilde{U}_\ell(B^p) = \bigoplus_{a+b=\ell+k-p} (\tilde{W}_\ell(B^p) \cap \tilde{F}^a(B^p) \cap \overline{\tilde{F}^b(B^p)}) .$$

The second statement in (4.4) guarantees that the differential d is homogeneous of degree -1 relative to this grading. Hence the purity statement would follow if we prove that every cocycle in $\tilde{U}_\ell(B^p)$ with $\ell > 0$ is exact. Now, $\tilde{U}_\ell(B^p)$ carries an indefinite, non-degenerate Hermitian form deduced from the Hodge structure (4.6) and the non-degenerate bilinear form $\tilde{S}_p(\cdot,\tilde{N}^\ell\cdot)$, $N = \underset{\sim}{N_1} + \ldots + N_n$. In what follows we shall fix p and denote this Hermitian form by $\tilde{H}_\ell$. The polarization statement in (4.4) implies that

(4.7) For $\ell \geq 0$, $\tilde{H}_\ell$ is positive definite on the primitive subspace

$$\tilde{U}_\ell(B^p) \cap \operatorname{Ker} \tilde{N}^{\ell+1} .$$

The adjoint δ of d, relative to $\tilde{H}_\ell$, is homogeneous of degree $+1$ with respect to the grading $\tilde{U}(B^p)$. We then have:

(4.8) For $\ell > 0$, $d\delta + \delta d\colon \tilde{U}_\ell(B^p) \longrightarrow \tilde{U}_\ell(B^p)$ is an isomorphism.

This statement is a consequence of (4.7) and the following two facts whose proof is contained in Lemma (3.12) of [CKS-2].

$$(4.9)\qquad \operatorname{Ker}\{d\delta + \delta d\colon \tilde{U}_\ell(B^p) \longrightarrow \tilde{U}_\ell(B^p)\} \subset (\tilde{U}_\ell(B^p) \cap \operatorname{Ker} \tilde{N}^{\ell+1}) .$$

(4.10) If $\omega \in \tilde{U}_\ell(B^p)$, $\ell > 0$, then $\tilde{H}_\ell((d\delta + \delta d)\omega,\omega) \geq \tilde{H}_\ell(\omega,\omega)$.

We conclude by showing that (4.8) implies a direct sum decomposition

$$(4.11)\qquad \tilde{U}_\ell(B^p) = \operatorname{Ker} \delta \oplus d(\tilde{U}_{\ell+1}(B^{p-1}))$$

from which it follows that every cocycle in $\tilde{U}_\ell(B^p)$ is cohomologous to one in $\operatorname{Ker} d \cap \operatorname{Ker} \delta = 0$ by (4.8). To prove (4.11), observe that the subspaces $\operatorname{Ker} \delta$ and $d(\tilde{U}_{\ell+1}(B^{p-1}))$ have complementary dimensions; on the other hand, if $\omega \in \operatorname{Ker} \delta \cap d(\tilde{U}_{\ell+1}(B^{p-1}))$ then $\omega = d\psi$, $\psi \in \tilde{U}_{\ell+1}(B^{p-1})$. By (4.8) applied to $\tilde{U}_{\ell+1}(B^{p-1})$, we can write $\psi = (d\delta + \delta d)\xi$ for some $\xi \in \tilde{U}_{\ell+1}(B^{p-1})$. Therefore:

$$0 = \delta(\omega) = \delta d(\psi) = \delta d((d\delta + \delta d)\xi) = \delta d\delta d(\xi) = (d\delta + \delta d)\delta d(\xi)$$

and appealing to (4.8) once more we obtain that $\delta d(\xi) = 0$ and consequently $\omega = d(d\delta + \delta d)\xi = 0$.

References

[CK-1] E. Cattani and A. Kaplan: Polarized mixed Hodge structures and the local monodromy of a variation of Hodge structures. Inventiones Math. 67, 101-115 (1982).

[CK-2] E. Cattani and A. Kaplan: On the SL(2)-orbits in Hodge Theory. IHES prepub./M/82/58. October 1982.

[CK-3] E. Cattani and A. Kaplan: Sur la cohomologie L_2 et la cohomologie d'intersection à coefficients dans une variation de structure de Hodge. C. R. Acad. Sc. Paris, 300, Série I, 351-353 (1985).

[CKS-1] E. Cattani, A. Kaplan and W. Schmid: Degeneration of Hodge structures, Ann. of Math., 123, 457-535 (1986).

[CKS-2] E. Cattani, A. Kaplan and W. Schmid: L_2 and intersection cohomologies for a polarized variation of Hodge structure. Inventiones Math. To appear.

[D] P. Deligne: Structures de Hodge mixtes réelles. Appendis to [CK-2].

[K] M. Kashiwara:The asymptotic behavior of a variation of polarized Hodge structure. Publ. RIMS, Kyoto Univ. 21, 853-875 (1985).

[Ko] J. Kollar: Subadditivity of the Kodaira dimension: fibers of general type. to appear.

[P] C. A. M. Peters: A criterion for flatness of Hodge bundles over curves and geometric applications. Math. Ann. 268, 1-19 (1984).

[S] W. Schmid: Variation of Hodge structure: the singularities of the period mapping. Inventiones Math. 22, 211-319 (1973).

SOME REMARKS ON L^2 AND INTERSECTION COHOMOLOGIES

Eduardo Cattani[1], Aroldo Kaplan[1] and Wilfried Schmid[2]
University of Massachusetts, Amherst, MA 01003
Harvard University, Cambridge, MA 02138

We recall Deligne's axiomatic definition of intersection cohomology [G-M,Bo]. Let $\bar{X}$ be a stratified topological space, $X \subset \bar{X}$ a dense open stratum, $\mathbf{V}$ a local system on X. Typically $\bar{X}$ arises as a complex analytic space, X as a Zariski open subset of the set of regular points, and $\bar{X}$ is stratified according to the degree of singularity along $\bar{X} - X$. For simplicity we assume that this is the case. Deligne's "perverse sheaf" $\mathbf{P}^*(\mathbf{V})$ is an extension of $\mathbf{V}$ to all of $\bar{X}$, not as a single sheaf, but rather in the class of bounded complexes of sheaves, and only up to quasi-isomorphism[3]; here $\mathbf{V}$ is viewed as a complex concentrated in degree zero. In addition to the extension property - i.e., $\mathbf{V}$ and $\mathbf{P}^*(\mathbf{V})$ are quasi-isomorphic over X - $\mathbf{P}^*(\mathbf{V})$ satisfies the following axioms:

(1)

a) its cohomology sheaves vanish in degree less than zero and are locally constant along the strata.

b) the cohomology sheaf in degree k, $k > 0$, is supported on the strata of complex codimension at least $k + 1$;

c) the inverse limit of $H_c^{2n-k}(U,\mathbf{P}^*(\mathbf{V}))$, as U ranges over the neighborhoods of a point x in a stratum S, with $n = \dim_{\mathbf{C}} X$ and $k > 0$, vanishes unless S has complex codimension at least $k + 1$.

($H_c^*(\ldots)$ denotes cohomology with compact support). These axioms characterize $\mathbf{P}^*(\mathbf{V})$ up to quasi-isomorphism. Any quasi-isomorphism of complexes of sheaves induces an isomorphism of their global (hyper-)cohomology groups, so $IH^*(\bar{X},\mathbf{V}) \underset{def}{=} H^*(\bar{X},\mathbf{P}^*(\mathbf{V}))$ is canonically attached to $\bar{X}$ and $\mathbf{V}$ - the intersection cohomology of $\bar{X}$ with values in $\mathbf{V}$, relative to the "middle perversity".

Intersection cohomology satisfies Poincaré duality: $IH^k(\bar{X},\mathbf{V})$ is naturally dual to $IH_c^{2n-k}(\bar{X},\mathbf{V}^*)$ $(= H_c^{2n-k}(\bar{X},\mathbf{P}^*(\mathbf{V}^*)))$; here $\mathbf{V}^*$ denotes the local system dual to $\mathbf{V}$.

---

1. Partially supported by NSF Grant DMS-8501949

2. Partially supported by NSF Grant DMS-8317436

3. A morphism between complexes of sheaves which induces an isomorphism of the associated cohomology sheaves is a particular kind of quasi-isomorphism; by definition, any quasi-isomorphism can be expressed as a composition of such morphisms and their formal inverses.

The deRham and Hodge theorems "explain" the classical statement of Poincaré duality. This has led to various conjectures relating intersection cohomology and L^2 cohomology [C-G-M,D,Z1]. The basic setting is the same in all cases: the fibres of $\mathbf{V}$ and the base X are endowed with hermitian metrics. One can then define a sheaf $\mathbf{A}^*_{(2)}(\mathbf{V})$ on $\bar{X}$, the sheaf of locally L^2 $\mathbf{V}$-valued differential forms, by the assignment

(2) $U \longrightarrow$ space of $\mathbf{V}$-valued forms ω on $U \cap X$, with locally L^2 coefficients, such that $d\omega$ exists as a locally L^2 form, with both ω and $d\omega$ globally L^2 on $K \cap X$, for any compact subset $K \subset U$.

In this definition, "locally L^2" specifies the degree of regularity of the coefficients, and is independent of the choice of metrics, whereas the global L^2 condition on subsets $K \cap X$ limits the growth of the coefficients along $\bar{X} - X$ in terms of the two metrics.

The standard Poincaré lemma applies at points of X, since X is a manifold. Thus $\mathbf{A}^*_{(2)}(\mathbf{V})$ and $\mathbf{V}$ are quasi-isomorphic over X. According to the conjectures, under appropriate hypotheses on $\bar{X}$, on the local system $\mathbf{V}$ and on the two metrics, $\mathbf{A}^*_{(2)}(\mathbf{V})$ also satisfies the axioms (1). In that case $\mathbf{A}^*_{(2)}(\mathbf{V})$ is quasi-isomorphic to $\mathbf{P}^*(\mathbf{V})$ over all of $\bar{X}$,

(3) $$\mathbf{A}^*_{(2)}(\mathbf{V}) \approx \mathbf{P}^*(\mathbf{V}), \text{ and hence } IH^*(\bar{X},\mathbf{V}) \cong H^*(\bar{X},\mathbf{A}^*_{(2)}(\mathbf{V})).$$

In certain situations - but not always - the sheaves $\mathbf{A}^q_{(2)}(\mathbf{V})$ are fine; the hypercohomology of $\mathbf{A}^*_{(2)}(\mathbf{V})$ can then be calculated in terms of the complex of global sections:

(4) $$H^*(\bar{X},\mathbf{A}^*_{(2)}(\mathbf{V})) \cong H^*(\Gamma\mathbf{A}^*_{(2)}(\mathbf{V})).$$

When $\bar{X}$ is compact, $\Gamma\mathbf{A}^*_{(2)}(\mathbf{V})$ consists precisely of the globally square-integrable, $\mathbf{V}$-valued forms ω on X, with locally L^2 coefficients, such that $d\omega$ is also globally L^2. If in addition to the compactness of $\bar{X}$, the metric of X is complete and the L^2 cohomology groups $H^*(\Gamma\mathbf{A}^*_{(2)}(\mathbf{V}))$ are finite dimensional, then the usual arguments of harmonic theory imply

(5) $$H^*(\Gamma\mathbf{A}^*_{(2)}(\mathbf{V})) \cong \text{space of } L^2 \text{ harmonic } \mathbf{V}\text{-valued forms on } X.$$

Integration over X pairs the L^2 harmonic spaces in complementary dimensions nondegenerately, so the isomorphisms (3-5) are compatible with Poincaré duality.

The identification (3) of intersection cohomology and L^2 cohomology has been conjectured in the following situations: i) for the trivial local system $\mathbf{C}$ over the set of regular points X in a complex projective variety $\bar{X}$, with the constant metric on $\mathbf{C}$ and the restriction to X of a metric on the ambient projective space

[C-G-M]; ii) for arithmetic quotients $X = \Gamma\backslash G/K$ of a hermitian symmetric space, lying in its Baily-Borel compactification $\bar{X}$, and any local system $\mathbf{V}$ coming from a representation of G, both equipped with quotients of G-invariant metrics [Z2]; iii) for local systems $\mathbf{V}$ which carry a polarizable variation of Hodge structure, over X, the complement of a divisor with normal crossings in a complex manifold $\bar{X}$ [D]; here $\mathbf{V}$ is endowed with the Hodge metric and X with a Kähler metric whose Kähler form is asymptotic, along the divisor $\bar{X} - X$, to the curvature form of that divisor.[4]

Let us suppose now that the local system $\mathbf{V}$ underlies a polarized variation of Hodge structure, that $\mathbf{V}$ is equipped with the Hodge metric, and that X carries a Kähler metric. There is a natural bigrading on the space of $\mathbf{V}$-valued differential forms which incorporates the Hodge bigradings on the fibres of $\mathbf{V}$ and on scalar valued forms. If in addition, $\bar{X}$ is assumed to be compact Kähler, Deligne (cf. [Z1]) has shown that the Kähler identities remain valid. In particular, the intersection cohomology groups $IH^*(\bar{X},\mathbf{V})$ inherit Hodge structures whenever the isomorphisms (3-5) are known to hold.

In the situation i), the isomorphism (3) has been established by Cheeger [C] for varieties with certain special types of isolated singularities; Hsiang-Pati [H-P] have extended Cheeger's argument to all normal surfaces $\bar{X}$. It is easy to see that the L^2 sheaves $\mathbf{A}^*_{(2)}(\mathbf{V})$ are fine in both cases, and this implies (4). The metric of the ambient projective space restricts to an incomplete metric on X, so the validity of (5) is far from clear - this seems to be a delicate analytic problem [C]. As it stands, the results of [H-P] do not produce a Hodge structure on the intersection cohomology. Hsiang-Pati and Saper [Sa] have circumvented this difficulty by working also with certain complete metrics on X. However, the metrics are not canonical in any reasonable sense, and the geometric significance of the resulting Hodge structures remains to be seen. Zucker and Borel-Casselman have proved (3) in the setting ii), for most hermitian symmetric spaces of rational rank one [Z2], respectively for all spaces of rational rank at most two [B-C]. The local system $\mathbf{V}$ underlies a natural variation of Hodge structure whenever the representation which determines $\mathbf{V}$ is defined over Q. In that case, since the metric on $X = \Gamma\backslash G/K$ is complete, the results of [Z2,B-C] do put Hodge structures on the intersection cohomology groups.

The isomorphism (3) in the situation iii), for variations of Hodge structure over curves, is due to Zucker [Z1]; in fact, Zucker's paper predates the definition of intersection cohomology. The same statement without restriction on the dimension of X was proved by us in [C-K-S2] and by Kashiwara and Kawai [K-K] (see also Kashiwara's article in this volume). We shall sketch our proof below. There are some

4. An explicit description of this type of metric will be given below. When $\bar{X}$ is compact Kähler, the existence of such a metric is automatic [C-G].

remarkable formal similarities between our arguments and those of [H-P,B-C]. When $\bar{X}$ is compact Kähler, both (4) and (5) can be verified easily in this setting, and consequently the intersection cohomology groups inherit canonical Hodge structures. The situation iii) differs from both i) and ii) in that $\bar{X}$ has no singularities at all; instead, the difficulties come from the behavior of the local system $\mathbf{V}$ near the divisor $\bar{X} - X$: the norm estimates and the purity theorem discussed in the companion paper [C-K-S3] are the most crucial ingredients of the proof of (3)[5].

It would be interesting to see if (3) and (5), in the situation iii), with $\bar{X}$ compact Kähler, remain valid when the complete metric on X is replaced with the incomplete metric obtained by restricting the metric of $\bar{X}$ to X. If so, there are two Hodge structures on the intersection cohomology, corresponding to the two choices of metrics; it should be possible to decide whether the two Hodge structures agree. By analogy, the answers to these questions might indicate if the incomplete metrics in the setting i) are likely to lead to geometrically significant Hodge structures.

We should also mention the recent work of Saito [S]. Saito puts Hodge structures on the intersection cohomology groups $IH^*(\bar{X},\mathbf{V})$ when $\mathbf{V}$ underlies a geometric variation of Hodge structure over a Zariski open subset X in a projective variety $\bar{X}$. He does this by formal reduction to the case of a one-dimensional variety $\bar{X}$; L^2 methods are used only at the final step of the reduction. There is considerable overlap between the situations i), ii), iii) and the setting of Saito's theorem. At present, it is not clear how the Hodge structures of [S] are related to those obtained via the isomorphisms (3-5).

We now consider the situation iii) in more detail. Thus, $\bar{X}$ is an n-dimensional complex manifold, $X \subset \bar{X}$ the complement of a divisor with normal crossings and $\mathbf{V} \to X$ a local system of $\mathbf{C}$-vector spaces. Every point $p \in \bar{X}$ has a neighborhood U such that

(6) $$U \cong \Delta^n;\quad U \cap X \cong (\Delta^*)^r \times \Delta^{n-r}\ .$$

By our assumptions, X carries a Kähler metric g whose restriction to such neighborhoods is quasi-isometric to a product of Euclidean metrics on the disk factors and metrics asymptotic - at the puncture - to the Poincaré metric on the Δ^*-factors. We suppose, moreover, that the local system $\mathbf{V}$ has quasi-unipotent monodromy locally around the divisor and that it underlies a polarized variation of Hodge structure of weight k. In particular, the Hodge metric defines a (non-flat) positive definite hermitian structure on $\mathbf{V}$.

---

5. We shall freely use both the notation and the results of [C-K-S3].

Theorem. The L^2-sheaves $\mathbf{A}^*_{(2)}(\mathbf{V})$ satisfy the axioms (1). Thus, $H^*_{(2)}(X,\mathbf{V}) \cong IH^*(\bar{X},\mathbf{V})$.

Corollary. When $\bar{X}$ is compact Kähler, the intersection cohomology groups $IH^p(\bar{X},\mathbf{V})$ carry canonical pure Hodge structures of weight $k + p$.

We point out that the Lefschetz decomposition and the Hodge-Riemann bilinear relations carry over to this setting.

The rest of this note will be devoted to a sketch of the proof of the Theorem. We will emphasize the main steps and refer the reader to [C-K-S2] for the details.

It suffices to prove a local isomorphism

(7) $$H^*_{(2)}(U,\mathbf{V}) \cong IH^*(U,\mathbf{V}).$$

Indeed, it is clear that $\mathbf{A}^*_{(2)}(\mathbf{V})$ satisfies (1a), while (7) together with Poincaré duality for L^2-cohomology (cf. (5.9) in [C-K-S2]) then imply that (1b-c) are satisfied by the L^2-complex. Moreover, one can see that it is enough to consider the case $r = n$, and to assume that the local monodromy of $\mathbf{V}$ is unipotent.

By induction on the length of the stratification we may assume that the theorem holds in the case $X = (\Delta^*)^n$ and $\bar{X} = (\Delta^n - \{0\})$. The exact cohomology sequence of a pair gives

$$\longrightarrow IH^p_c(\Delta^n,\mathbf{V}) \longrightarrow IH^p(\Delta^n,\mathbf{V}) \longrightarrow IH^p(\Delta^n - \{0\},\mathbf{V}) \longrightarrow IH^{p+1}_c(\Delta^n,\mathbf{V}) \longrightarrow$$

but, on the other hand, (1,b-c) applied to $IH^*(\Delta^n,\mathbf{V})$ mean that

(8) $$IH^p(\Delta^n,\mathbf{V}) = 0 \text{ if } p \geq n; \quad IH^p_c(\Delta^n,\mathbf{V}) = 0 \text{ if } p \geq n .$$

Consequently,

(9) $$IH^p(\Delta^n,\mathbf{V}) \cong IH^p(\Delta^n - \{0\},\mathbf{V}) \text{ if } p < n ,$$

and appealing also to Poincaré duality for intersection cohomology

(10) $$IH^p(\Delta^n - \{0\},\mathbf{V}) \cong IH^{p+1}_c(\Delta^n,\mathbf{V}) \cong IH^{2n-p-1}(\Delta^n,\mathbf{V})^* \text{ for } p \geq n .$$

In view of (8), the isomorphism (9) reduces the proof of (7) to the following L^2-cohomology statements:

(11) $$H^p_{(2)}(\Delta^n - \{0\},\mathbf{V}) \cong H^p_{(2)}(\Delta^n,\mathbf{V}) \quad \text{if } p < n \quad \text{and}$$
$$H^p_{(2)}(\Delta^n,\mathbf{V}) = 0 \quad \text{if } p \geq n.$$

Recall the complex (B^*,d) defined in §4 of [C-K-S3].

Proposition. (Deligne). The complex (B^*,d) computes the local intersection cohomology $IH^*(\Delta^n,\mathbf{V})$.

We recall also that each B^p carries a filtration $\tilde{W}(B^p)$ which, after an appropriate shifting, is the weight filtration of a mixed Hodge structure split over $\mathbf{R}$. This gives a canonical $\mathbf{R}$-grading of the filtration $\tilde{W}(B^p)[-p]$, preserved by the differential (cf. (4.4) in [C-K-S3]). Thus we obtain a grading of the intersection cohomology $IH^p(\Delta^n,\mathbf{V}) = \bigoplus_{\ell\geq 0} (IH^p(\Delta^n,\mathbf{V}))_\ell$ and the purity theorem ([C-K-S3], (4.5)) gives

$$(12) \qquad (IH^p(\Delta^n,\mathbf{V}))_\ell = 0 \qquad \text{for } \ell > p .$$

The isomorphisms (9-10) allow us to grade $IH^*(\Delta^n - \{0\},\mathbf{V})$ as well; we set

$$(13) \qquad \begin{aligned} (IH^p(\Delta^n - \{0\},\mathbf{V}))_\ell &\cong (IH^p(\Delta^n,\mathbf{V}))_\ell && \text{if } p < n, \\ (IH^p(\Delta^n - \{0\},\mathbf{V}))_\ell &\cong ((IH^{2n-p-1}(\Delta^n,\mathbf{V}))_{2n-\ell})^* && \text{if } p \geq n . \end{aligned}$$

Combining (12) and (13) we obtain

$$(14) \qquad (IH^p(\Delta^n - \{0\},\mathbf{V}))_\ell = 0 \text{ if either } p < n \leq \ell \text{ or } \ell \leq n \leq p.$$

In particular $(IH^*(\Delta^n - \{0\},\mathbf{V}))_n = 0$.

We will next define gradings in the L^2-cohomologies $H^*_{(2)}(D,\mathbf{V})$, where we let D stand for Δ^n or $\Delta^n - \{0\}$ (or indeed, any rotation-invariant open subset of Δ^n). After showing that the gradings on $H^*_{(2)}(\Delta^n - \{0\},\mathbf{V})$ and $IH^*(\Delta^n - \{0\},\mathbf{V})$ are compatible with the inductive hypothesis, we will be able to carry the vanishing statements (14) over to L^2-cohomology. It is in this way that purity enters into the proof of (11).

We observe that, since the sheaves $\mathbf{A}^*_{(2)}(\mathbf{V})$ are fine, the complex of global sections $\Gamma(D,\mathbf{A}^*_{(2)}(\mathbf{V}))$ computes the cohomology[6] $H^*_{(2)}(D,\mathbf{V})$. Lifting to U^n - the product of upper half-planes - elements of $(D,\mathbf{A}^*_{(2)}(\mathbf{V}))$ may be expressed as linear combinations of terms of the form

$$(15) \qquad \psi(\underline{z}) \otimes v(\underline{z}) \; ; \quad \tilde{v}(\underline{z}) = \exp(\sum_{1\leq j\leq n} x_j N_j)v$$

---

[6] The elements of $(\Delta^n,\mathbf{A}^*_{(2)}(\mathbf{V}))$ do not satisfy global L^2 conditions; however, one can prove (cf. [C-K-S2], (4.30)) that the cohomology $H^*_{(2)}(\Delta^n,\mathbf{V})$ may be computed by the complex of forms that are globally square integrable and have globally square integrable exterior derivatives. These conditions are taken relative to the metric g of X, which is asymptotic to the Poincare metric in $(\Delta^*)^n$ but extends across the boundary of Δ^n.

where: $\underline{z} = (z_1,\ldots,z_n) \in \mathcal{U}^n$, $z_j = x_j + iy_j$, ψ is a differential form on $\mathcal{U}^n$ whose coefficients are locally L^2 and which has a locally L^2 exterior derivative, v is an element of V (= typical fibre of $\mathbf{V}$) viewed as a flat section of the pullback of $\mathbf{V}$ to $\mathcal{U}^n$; consequently, $\tilde{v}(\underline{z})$ descends to a section of $C^\infty \otimes \mathbf{V}$ over $(\Delta^*)^n$. We let $\mathbf{R}^n$ act on the terms (15) by real translations on the coefficients of ψ and dropping to $(\Delta^*)^n$ we obtain an action of the torus T^n on $\Gamma(D,\mathbf{A}^*_{(2)}(\mathbf{V}))$; we have

Lemma. The complex of T^n-invariant global sections of $\mathbf{A}^*_{(2)}(\mathbf{V})$ over D, computes the cohomology $H^*_{(2)}(D,\mathbf{V})$.

This result is standard in the absence of monodromy; to prove it in our case we filter $\mathbf{V}$ by the weight filtration $\mathbf{W}$ of the monodromy cone C. This gives complexes $\mathbf{A}^*_{(2)}(\mathbf{W}_\ell)$ and $\mathbf{A}^*_{(2)}(Gr_\ell\mathbf{W})$ defined relative to the induced Hodge metric. The cohomology of the latter complex may then be computed by T^n-invariant global sections. The assertion now follows by considering the spectral sequence of the filtration $\mathbf{W}$ after observing that, because of the norm estimates ([C-K-S3], (3.7)), we have an exact sequence of complexes

$$0 \longrightarrow \Gamma(D,\mathbf{A}^*_{(2)}(\mathbf{W}_{\ell-1})) \longrightarrow \Gamma(D,\mathbf{A}^*_{(2)}(\mathbf{W}_\ell)) \longrightarrow \Gamma(D,\mathbf{A}^*_{(2)}(Gr_\ell\mathbf{W})) \longrightarrow 0 \quad .$$

Consequently, in (15) it suffices to consider expressions of the form

$$(16) \qquad f(\underline{y})dy_{\underline{I}} \wedge dx_{\underline{J}} \otimes \tilde{v}(\underline{z});$$

where $I,J \subset \{1,\ldots,n\}$, $f(\underline{y})$ is a function on the positive quadrant $\{\underline{y} = (y_1,\ldots,y_n): y_j > 0\}$ satisfying the appropriate regularity conditions and $dy_{\underline{I}}$ (resp. $dx_{\underline{J}}$) stands for the wedge product of dy_i, $i \in I$ (resp. dx_j, $j \in J$). We let $V = \bigoplus_{-k\le\ell\le k} V_\ell$ denote an $\mathbf{R}$-grading of the weight filtration W of the monodromy cone C, compatible with the N_i's, i.e. $N_iV_\ell \subset V_{\ell-2}$. Define a semisimple transformation Y acting on the T^n-invariant sections of $\mathbf{A}^*_{(2)}(\mathbf{V})$ over D, by letting Y act on an element of the form (16) with $v \in V_\ell$, as multiplication by $\ell + 2|J|$. Since the differential of $\tilde{v}(\underline{z})$ is given by

$$\sum_{1\le j\le n} N_jv \otimes dx_j \ ,$$

it follows that the action of Y commutes with the differential. On the other hand, the pull-back to $\mathcal{U}^n$ of a T^n-invariant form $\phi \in \Gamma(D,\mathbf{A}^*_{(2)}(\mathbf{V}))$, when restricted to a region as in (3.6) of [C-K-S3], may be written as a linear combination of terms of the form

$$f(\underline{y})dy_{\underline{I}} \wedge dx_{\underline{J}} \otimes \tilde{v}_j(\underline{z})$$

where the $\tilde{v}_j(\underline{z})$ define an L^2-adapted frame. The norm estimates ([C-K-S3], (3.7)) now imply that the action of Y preserves the L^2-conditions. In view of the Lemma, Y defines a grading in $H^*_{(2)}(D,\mathbf{V})$.

<u>Proposition</u>. The isomorphism $IH^*(\Delta^n - \{0\},\mathbf{V}) \cong H^*_{(2)}(\Delta^n - \{0\},\mathbf{V})$ is compatible with the gradings.

In order to prove this proposition, one shows that the gradings on $H^p_{(2)}(\Delta^n - \{0\},\mathbf{V})$ and $IH^p(\Delta^n,\mathbf{V})$ arise from maps of the local system $\mathbf{V}$ which, restricted to $(\Delta^*)^n$, extend continuously to Δ^n and preserve the natural stratification. These two maps are related by a stratification-preserving homeomorphism of Δ^n. From (13) one can then obtain that the statement holds for $p < n$. A careful application of Poincaré duality yields the statement for $p \geq n$.

To complete the proof of (11) we show that the cohomologies $H^*_{(2)}(\Delta^n,\mathbf{V})$ and $H^*_{(2)}(\Delta^n - \{0\},\mathbf{V})$ are related via a Mayer-Vietoris spectral sequence. We fix $r > 1$ and cover the positive quadrant with the open sets

$$Q_j(r) = \{\underline{y} \in \mathbf{R}^n : y_i > 0,\ ry_i > y_j \quad \text{for} \quad i \neq j\} .$$

Intersections of the Q_j's correspond to sets of the form

$$Q_J(r) = \bigcap_{j \in J} Q_j(r) \qquad \text{for} \qquad J = \{j_1,\dots,j_p\} \subset \{1,\dots,n\} .$$

We will denote by $D_J(r)$ the projection of $\mathbf{R}^n + iQ_J(r)$ to $(\Delta^*)^n$, and by $A^*(D_J) = A^*(D_J(r),\mathbf{V})$ (respectively $\tilde{A}^*(D_J) = \tilde{A}^*(D_J(r),\mathbf{V})$) the complex of T^n-invariant, $\mathbf{V}$-valued forms on $D_J(r)$, satisfying the appropriate regularity conditions and which are square-integrable over $K \cap D_J(r)$ for every compact subset $K \subset \Delta^n$ (respectively $K \subset \Delta^n - \{0\}$). A partition of unity argument shows that there exist Mayer-Vietoris spectral sequences which abut to $H^*_{(2)}(\Delta^n,\mathbf{V})$ (respectively $H^*_{(2)}(\Delta^n - \{0\},\mathbf{V})$) and whose E_1-terms are

$$E_1^{p,q} = \bigoplus_{|J|=p+1} H^q(A^*(D_J)) \quad (\text{respectively } \tilde{E}_1^{p,q} = \bigoplus_{|J|=p+1} H^q(\tilde{A}^*(D_J))) .$$

It now suffices to prove (11) at the E_1-level; to relate the cohomologies $H^q(A^*(D_J))$ and $H^q(\tilde{A}^*(D_J))$ we assume that $J = \{1,\dots,p+1\}$ and introduce new variables on $Q_J(r)$ as follows:

$$t = y_1 \ ; \quad u_i = \frac{y_i}{y_1} , \quad 2 \leq i \leq p+1; \qquad v_j = \frac{y_j}{y_1} , \quad p+1 < j \leq n .$$

Given a section $\tilde{v}(\underline{z})$ as in (15), with v in V_ℓ, the estimate (3.7) of [C-K-S3] shows that

$$\|v\|_{(\underline{x},t,\underline{u},\underline{v})} \sim t^{\ell/2}\|v\|_{(\underline{x},1,\underline{u},\underline{v})} \quad .$$

If we now define S_J as the intersection of Q_J with the hyperplane $\{t = 1\}$, so that $Q_J = \mathbf{R}^+ \times S_J$, then a variant of the Künneth formula allows us to compute the two cohomologies $H^q(A^*(D_J))$ and $H^q(\tilde{A}^*(D_J))$ in terms of the cohomology of an appropriate complex constructed over S_J and weighted (by weights which depend on ℓ) L^2-cohomologies over $\mathbf{R}^+$. For these, the L_2-conditions on $A^*(D_J)$ translate into square-integrability over intervals (a,∞), while those for $A^*(D_J)$ become essentially vacuous. One thus obtains isomorphisms

$$(17)\qquad \begin{aligned} H^q(A^*(D_J))_\ell &\cong H^{q-1}(\tilde{A}^*(D_J))_\ell && \text{if } \ell < n, \\ &\cong H^{q-1}(\tilde{A}^*(D_J))_n \oplus M && \text{if } \ell = n, \\ &\cong 0 && \text{if } \ell > n \quad , \end{aligned}$$

where M is an infinite dimensional vector space given by

$$M = \frac{\{f(t) : t^{1/2}f(t) \in L^2([1,\infty))\}}{\{f'(t) : t^{-1/2}f(t), t^{1/2}f'(t) \in L^2([1,\infty))\}}$$

The isomorphisms (17), together with the vanishing statements (14) now imply (11).

References

[Bo] A. Borel et al.: Intersection cohomology. Progress in Math. 50 (1984), Birkhäuser, Boston.

[B-C] A. Borel and W. Casselman: Cohomologie d'intersection et L^2-cohomologie de variétés arithmétiques de rang rationnel 2. C. R. Acad. Sc. Paris, 301, Série I, 369-373 (1985).

[C-K-S1] E. Cattani, A. Kaplan and W. Schmid: Degeneration of Hodge structures. Ann. of Math., 123, 457-535 (1986).

[C-K-S2] E. Cattani, A. Kaplan and W. Schmid: L_2 and intersection cohomologies for a polarized variation of Hodge structure. Invent. math. To appear.

[C-K-S3] E. Cattani, A. Kaplan and W. Schmid: Variations of polarized Hodge structure: asymptotics and monodromy. This volume.

[C] J. Cheeger: On the Hodge theory of Riemannian pseudomanifolds. Proc. Symp. Pure Math. AMS 36, 91-146 (1980).

[C-G-M] J. Cheeger, M. Goresky and R. MacPherson: L^2-cohomology and intersection homology for singular algebraic varieties. In: Seminar on Differential Geometry, S.-T. Yau, ed., 303-340. Princeton Univ. Press, (1982).

[C-G] M. Cornalba and P. Griffiths: Analytic cycles and vector bundles on non compact algebraic varieties. Invent. math. 28, 1-106 (1975).

[D] P. Deligne: Personal communication, dated January 16, 1981.

[G-M] M. Goresky and R. MacPherson: Intersection homology, II, Invent. math. 71, 77-129 (1983).

[H-P] W. -C. Hsiang and V. Pati: L^2-cohomology of normal algebraic surfaces, I. Invent. math. 81, 395-412 (1985).

[K] M. Kashiwara: Poincaré lemma for a variation of polarized Hodge structure. This volume.

[K-K] M. Kashiwara and T. Kawai: The Poincaré lemma for a variation of polarized Hodge structure. Proc. Japan Acad., 61, 164-167 (1985).

[S] M. Saito: Modules de Hodge polarisables. Preprint.

[Sa] L. Saper: L_2-cohomology and intersection homology of certain algebraic varieties with isolated singularities. Invent. math. 82, 207-255 (1985).

[Z1] S. Zucker: Hodge theory with degenerating coefficients: L_2-cohomology in the Poincaré metric. Ann. of Math. 109, 415-476 (1979).

[Z2] S. Zucker: L_2-cohomology of warped products and arithmetic groups. Invent. math. 70, 169-218 (1982).

THE L-ADIC COHOMOLOGY OF LINKS

Alan H. Durfee
Mount Holyoke College
South Hadley, MA 01075

Let X be a complex projective algebraic variety, and let E be a closed subvariety with the singular locus of X contained in E. The link L of E in X is defined to be the boundary of a small tubular neighborhood of E in X. Previous papers [Elzein], [Durfee], [Guillen et al.], [Steenbrink] have shown that the rational cohomology of the link L has a mixed Hodge structure. This mixed Hodge structure consists of a weight and Hodge filtration satisfying various properties. Recent work [Durfee & Hain] (and Navarro, unpublished) has shown that the real homotopy type of L has a mixed Hodge structure. A consequence of this result is that the weight filtration on the cohomology of L is preserved by the cup product. This was used to find new restrictions on the topology of links.

The purpose of this note is to describe the cohomology of the link as an ℓ-adic group, and to describe the weight filtration using the Frobenius automorphism. In particular, this note gives an easy proof that the cup product in cohomology preserves the weight filtration. (The proof will actually work for the more general situation of the complement of one link in another, as in [Durfee & Hain].)

An outline of a proof of this type in the case of varieties (rather than links) is sketched in [Deligne, Vancouver]. This proof proceeds along the same lines; the general idea was suggested to the author by Deligne.

We use $D(R)$ for the derived category of (bounded) R-modules, and $DF(R)$ for the corresponding derived filtered category.

1. The cohomology and weight filtration of the link via mixed Hodge theory

The link L as defined above is a homotopy class of spaces, and hence has a well-defined integral cohomology $H^k_{an}(L)$, for all k. This cohomology has a mixed Hodge structure; this consists of an ascending weight filtration W and a descending Hodge filtration satisfying various requirements. We will only use the weight filtration in this paper. The mixed Hodge structure comes from a mixed Hodge complex; this consists of an object of the filtered derived category $DF(\mathbb{Q})$, and an object of the bifiltered derived category $DF_2(\mathbb{C})$. We will only work with the object in $DF(\mathbb{Q})$. The homology of this object is isomorphic to $H_{an}(L;\mathbb{Q})$, and the filtration of this object induces the filtration W, with the usual shift in indices. We now describe how this mixed Hodge complex is constructed.

We may assume without loss of generality that X is smooth projective and that E is a divisor with normal crossings. Let $E = E^1 \cup \ldots \cup E^k$ be its decomposition into irreducible components. Let $U = X-E$, and let $i: E \to X$ and $j: U \to X$ be the inclusion maps . Suppose that A is an abelian group. Let

$$S(A) = i^* Rj_* A_U$$

where A_U is the constant sheaf A on U. It is not hard to show that

$$H^k_{an}(L;A) \cong R^k\Gamma_{an} S(A)$$

where Γ_{an} is the analytic global section functor.

The general reference for the following is [Deligne, Hodge III, §5,7]. Let $E_\cdot = \mathrm{cosk}(\coprod E^i)$. Then $a: E_\cdot \to E$ is a simplicial resolution of E, and

$$R\Gamma_{an} S(A) \cong R\Gamma_{an} a^* S(A)$$

in $D(\mathbb{Z})$, so in particular, $R^k\Gamma_{an}a^*S(A) \cong H^k_{an}(L;A)$. The object $R\Gamma_{an}a^*S(A)$ has an increasing filtration $\delta(\tau,L)$, the diagonal filtration δ of the canonical filtration τ on Rj_*A_U and the simplicial filtration L on $R\Gamma_{an}a^*S(A)$. Let W' be the filtration $\delta(\tau,L)$. The spectral sequence of the decreasing filtration corresponding to W' converges to $H_{an}(L;A)$. When $A = \mathbb{Q}$, this spectral sequence degenerates at the E_2 term, and $E_2^{\cdot s} \cong Gr^W_s[H(L;\mathbb{Q})]$. The object in $DF(\mathbb{Q})$ giving the mixed Hodge complex for L is defined to be $(R\Gamma_{an}a^*S(\mathbb{Q}), W')$.

It is not hard to show that $(R\Gamma_{an}a^*S(\mathbb{Q}), W')\otimes\mathbb{R}$ is isomorphic to the real mixed Hodge complex for L constructed in [Durfee & Hain].

2. The ℓ-adic cohomology of the link

Now suppose that the abelian group A is finite. The comparison theorem [Deligne, SGA 4 1/2 p.51] says that there are isomorphisms

$$R\Gamma_{an}S(A) \cong R\Gamma S(A)$$

and

$$R\Gamma_{an}a^*S(A) \cong R\Gamma a^*S(A) \qquad (1)$$

where Γ is the étale global section functor. The filtration $W'=\delta(\tau,L)$ can be defined on $R\Gamma a^*S(A)$ exactly the same way it is defined on $R\Gamma_{an}a^*S(A)$. Hence the isomorphism (1) holds in the derived filtered category $DF(\mathbb{Z})$ as well. In particular, the E_1 terms of the spectral sequences converging to $H(R\Gamma_{an}a^*S(A))$ and $H(R\Gamma a^*S(A))$ are isomorphic.

Let ℓ be a prime. The ℓ-adic cohomology of the link L is defined, for all k, by

$$H^k(L;\mathbb{Q}_\ell) = \left[\lim_{\leftarrow} R^k\Gamma S(\mathbb{Z}/\ell^n\mathbb{Z})\right]\otimes\mathbb{Q}_\ell$$

Since

$$R^k\Gamma_{an}S(A) \cong R^k\Gamma S(A)$$

for all finite abelian groups A, and $H^k(L)$ is finitely generated, we have that

$$H^k_{an}(L;\mathbb{Q}_\ell) \cong H^k(L;\mathbb{Q}_\ell)$$

Similarly,

$$\left[\lim_{\leftarrow} R^k\Gamma_{an}a^*S(\mathbb{Z}/\ell^n\mathbb{Z})\right]\otimes\mathbb{Q}_\ell \cong \left[\lim_{\leftarrow} R^k\Gamma a^*S(\mathbb{Z}/\ell^n Z)\right]\otimes\mathbb{Q}_\ell$$

and both are isomorphic to the above groups.

Since the E_1 terms of the spectral sequences converging to $R^k\Gamma_{an}a^*S(\mathbb{Z}/\ell^n\mathbb{Z})$ and $R^k\Gamma a^*S(\mathbb{Z}/\ell^n\mathbb{Z})$ are isomorphic, and since the inverse limit of the first spectral sequence gives a spectral sequence converging to $H_{an}(L;\mathbb{Q}_\ell)$, the inverse limit of the second spectral sequences gives a spectral sequence converging to $H(L;\mathbb{Q}_\ell)$. Furthermore, the E_1 terms of both spectral sequences are isomorphic.

3. The ℓ-adic grading of the weight filtration

There is a ring $R \subset \mathbb{C}$ which is finitely generated over the integers such that X, U and the simplicial space $E.$ are defined over R (that is, they are obtained from the corresponding schemes over R by extension of scalars). Let $K \subset \mathbb{C}$ be the field of fractions of R. By inverting more primes we may assume without loss of generality that the simplicial space $E.$ is smooth over R.

Let ℓ be a prime number. Choose a maximal ideal m in $R[1/\ell]$, and let $k = R/m$; then k is a finite field with q elements, say, and q and ℓ are relatively prime. Let $\bar{k}$ be an algebraic closure of k. The group $Gal(\bar{k}/k)$ is generated by a Frobenius element ϕ_m, where $\bar{\phi}_m(x) = x^q$. There is a conjugacy class of automorphisms in $Gal(K^u/K)$ corresponding to $\bar{\phi}_m$, where K^u is the maximal extension of K in its algebraic closure in $\mathbb{C}$ unramified at $R[1/\ell]$ [Serre, §I.8]. Extend these arbitrarily to a class of automorphisms ϕ_m in $Gal(\mathbb{C}/K)$. The maps $F_m = \phi_m^{-1}$ are the geometric Frobenius. The action induced by this on the ℓ-adic cohomology of a variety over R is well defined.

The weight filtration W on the rational cohomology of L carries over to ℓ-adic cohomology by means of the isomorphism $H^k(L;\mathbb{Q}_\ell) \cong H^k_{an}(L;\mathbb{Q})\otimes\mathbb{Q}_\ell$. Fix k. Define subspaces $\overline{W}_s$ of $H^k(L;\mathbb{Q}_\ell)$ as follows: The Galois automorphism F_m of $\mathbb{C}/K$ induces a well-defined automorphism F_m^* of $H^k(L;\mathbb{Q}_\ell)$. Let

$$\overline{W}_s = \left\{ \begin{array}{l} v \in H^k(L;\mathbb{Q}_\ell): \quad v \text{ is in a generalized eigenspace} \\ \text{of } F_m^* \text{ with eigenvalue of absolute value } q^{s/2} \end{array} \right\}$$

It will turn out that the subspaces $\overline{W}_s$ form a splitting for the filtration W, i.e. that $W_m = \oplus_{s \leqq m}\overline{W}_s$.

First suppose that $r \geqq 0$. The automorphism F_m^* acts on the spectral sequence converging to $H(L;\mathbb{Q}_\ell)$. A computation shows that $E_1^{rs} \cong H^s(E_r \times_R \mathbb{C};\mathbb{Q}_\ell)$, where E_r is the r-th component of the simplicial space E. This component is smooth and projective over R. By the proper and smooth base change theorems [Milne p. 230], there is an isomorphism

$$H(E_r \times_R \mathbb{C};\mathbb{Q}_\ell) \cong H(E_r \times_R \overline{k};\mathbb{Q}_\ell)$$

This isomorphism is equivariant with respect to the action of the geometric Frobenius. By [Deligne, Weil I], the eigenvalues of the action on $H^s(E_r \times_R \overline{k};\mathbb{Q}_\ell)$ are algebraic integers of absolute value $q^{s/2}$. Hence the same is true for $H^s(E_r \times_R \mathbb{C};\mathbb{Q}_\ell)$ and hence E_1^{rs}. Since E_2^{rs} is a subquotient, it is true for this as well. But $E_2^{rs} \cong Gr_s^W H^{r+s}(L;\mathbb{Q}_\ell)$, so $Gr_s^W H^{r+s}(L;\mathbb{Q}_\ell) \cong \overline{W}_s H^{r+s}(L;\mathbb{Q}_\ell)$. A similar argument works for $r<0$. Hence the subspaces $\overline{W}_s$ form a splitting for W as claimed above.

4. The cup product

Finally, let us prove that the cup product preserves the weight filtration. The map F_m^* is a geometric and hence preserves the cup product in $H(L;\mathbb{Q}_\ell)$. Clearly $\bar{W}_r \cup \bar{W}_s \subset \bar{W}_{r+s}$. Hence $W_r \cup W_s \subset W_{r+s}$ in $H(L;\mathbb{Q}_\ell)$. Since $H(L;\mathbb{Q}_\ell) \cong H_{an}(L;\mathbb{Q})\otimes\mathbb{Q}_\ell$, the cup product in $H_{an}(L;\mathbb{Q})$ preserves the weight filtration as well.

Acknowledgments: I would like to thank D. Cox for comments on a preliminary version of this paper.

REFERENCES

P. Deligne, Poids dans la cohomologie des variétés algébrique, Acts du Congrès International des Mathématiciens (Vancouver, 1974) Canadian Math Congress, 1975, I,79-85.

---, Théorie de Hodge III, Publications Math de IHES, 44 (1975),5-77.

---, La conjecture de Weil I, Publications Math de IHES (1974), 273-308.

---, Séminaire de Géométrie Algébraique du Bois-Marie SGA 4 1/2, Lecture Notes in Mathematics 569, Springer 1977.

A. Durfee, Mixed Hodge structures on punctured neighborhoods, Duke Math. J. 50 (1983), 1017-1040.

A. Durfee and R. Hain, Mixed Hodge structures on the homotopy of links, preprint, October 1985.

F. Elzein, Mixed Hodge structures, Trans. A.M.S. 275 (1983), 71-106.

F. Guillen, V. Aznar and F. Puerta, Théorie de Hodge via schemas cubique, preprint, June 1982, Universidad Politecnica de Barcelona.

J. Milne, Etale Cohomology, Princeton Univ. Press 1980.

J.-P. Serre, Local Fields, Springer 1979.

J. Steenbrink, Mixed Hodge structures associated with isolated singularities, Proc. Sym. Pure Math., 40 Part 2 (1983), 513-536.

HYPERRÉSOLUTIONS CUBIQUES ET APPLICATIONS À LA THÉORIE DE HODGE-DELIGNE

F. Guillén, F. Puerta
Dept. de Matemàtiques, ETSEIB, Universitat Politècnica de Catalunya, Diagonal 647, Barcelona 08028, Espagne

INTRODUCTION

Soit X une variété algébrique complexe. Dans [3] Deligne a développé une théorie qui munit les groupes de cohomologie $H^*(X, Z)$ d'une structure de Hodge mixte fonctorielle, qui généralise la théorie de Hodge des variétés projectives lisses.

La théorie de Deligne se développe en deux étapes. Dans la première on étudie le cas d'une variété X lisse, en considérant une compactification lisse $\bar{X}$ de X dans laquelle $\bar{X}$-X soit un diviseur a croisements normaux. Dans la seconde sa méthode de descente cohomologique simpliciale permet de ramener le cas général, où X est possiblement singulière et non compacte, au cas précédent en substituant X par un schéma simplicial $X_{\bullet}$ qui est le complémentaire d'un diviseur à croisements normaux dans un schéma simplicial propre et lisse. Dans cette deuxième étape, la technique des hyperrésolutions cubiques de V. Navarro Aznar fournit une méthode alternative au méthode simpliciale qui permet, de façon analogue, de remplacer le schéma X par un schéma cubique $X_{\bullet}$ qui est le complémentaire d'un diviseur à croisements normaux dans un schéma cubique propre et lisse qui satisfait en plus la relation

$$\dim X_n \leq \dim X - n .$$

Le développement systématique de la théorie de Hodge-Deligne du point de vue cubique se trouve dans [9] qui est une version revisée et augmentée de [8]. Ce travail contient des applications des hyperrésolutions cubiques à d'autres problèmes formulés en termes de théories cohomologiques dont on connait leur solution dans certains cas particuliers, concrétement aux théorèmes sur la monodromie, à la cohomologie de De Rham algébrique, aux théorèmes d'annulation de Kodaira-Akizuki-Nakano et de Grauert-Riemenschneider et à la théorie K algébrique supérieure et théorèmes de Riemann-Roch.

Dans ce travail nous nous bornons à montrer quelques applications des hyperrésolutions cubiques à la théorie de Hodge-Deligne qui se dérivent de la finitude et du contrôle des dimensions signalés précédemment. Nous allons décrire le contenu du travail.

Dans le § I on expose la construction des hyperrésolutions cubiques. Cette construction suit un processus récurrent et essentiellement combinatoire basé sur les théorèmes d'Hironaka et sur des propriétés élémentaires des modifications propres. Dans ce processus la dimension des schémas qui interviennent décroit à chaque étape, ce qui permet de s'arrêter après un nombre fini d'étapes. Les aspects cohomologiques sont aisément abordés avec la technique des complexes multiples. Ce fait s'illustre avec l'exemple suivant. Soit $\pi: \tilde{X} \longrightarrow X$ un morphisme propre de variétés qui est un isomorphisme en déhors d'une sous-variété fermée Y de X, et considérons le carré cartésien associé suivant

$$\begin{array}{ccc} \tilde{Y} & \longrightarrow & \tilde{X} \\ \downarrow & & \downarrow \pi \\ Y & \longrightarrow & X . \end{array}$$

Ce diagramme induit une suite exacte longue de cohomologie entière,

$$\ldots \to H^i(X) \to H^i(Y) \oplus H^i(\tilde{X}) \to H^i(\tilde{Y}) \to H^{i+1}(X) \to \ldots ,$$

qui peut s'obtenir à partir de la propriété locale suivante: le diagramme des complexes de faisceaux sur X,

$$\begin{array}{ccc} \mathbb{R}\pi_* Z_{\tilde{Y}} & \longleftarrow & \mathbb{R}\pi_* Z_{\tilde{X}} \\ \uparrow & & \uparrow \\ Z_Y & \longleftarrow & Z_X , \end{array}$$

considéré comme complexe multiple, est tel que le complexe simple associé est acyclique.

L'utilisation des hyperrésolutions cubiques pour l'obtention de structures de Hodge mixtes s'applique dans plusieurs contextes:

a) Cohomologie et homologie d'un schéma, et d'un schéma cubique,

b) Cohomologie locale, $H^*_Y(X, Z)$, et cohomologie du noeud, $H^*(X-Y, Z)$, d'un espace analytique X qui se rétracte sur une sous-variété algébrique compacte Y .

c) Cohomologie de la fibre limite d'une famille, $f: X \longrightarrow \mathbb{D}$, de variétés algébriques compactes parametrisée par le disque $\mathbb{D}$.

Ces applications s'exposent dans les §§ II, III et IV.

Dans le cas a) la structure de Hodge mixte s'obtient par une méthode entièrement analogue à celle de Deligne. Les hyperrésolutions cubiques donnent des bornes pour la dimension du support des termes de l'espace gradué associé à la filtration par le poids à niveau de faisceaux, qui confirment la relation conjecturée par McCrory [10] entre la filtration par le poids de Deligne et la filtration de Zeeman dans la cohomologie rationnelle d'une variété algébrique compacte. Cette conjecture est prouvée dans le § III avec la théorie des courants [7].

Le cas b) se ramène à la situation dans laquelle X est lisse et Y est un diviseur à croisements normaux dans X , situation qui a été étudiée par Fujiki [6], (cf [5]).

Au moyen d'un changement de base pour rendre unipotente la monodromie, le cas c) se ramène à la situation dans laquelle X est lisse, f est propre et lisse en déhors de l'origine $0 \in \mathbb{D}$, et $f^{-1}(0)$ est un diviseur à croisements normaux dans X . Cette situation a été étudiée auparavant par Schmid [13] et par Clemens [2] et Steenbrink [14]. En specialisant la suite spectrale associée à la filtration par l'indice cubique dans un point générique $t \in \mathbb{D}-0$ on obtient la dégénération de cette suite spectrale, ce qui montre aussitôt que la filtration par le poids de la fibre limite coïncide avec la filtration par la monodromie.

Les résultats des §§ I et IV sont dus à V. Navarro Aznar. Lui même a suggeré les questions tratées dans les §§ II et III, au deuxième et premier auteur respectivement, ainsi que les éléments nécessaires pour leur résolution. Nous lui en remercions très sincérement. Nous remercions aussi P. Pascual avec qui nous avons eu beaucoup de discussions sur le sujet.

§ I. HYPERRESOLUTIONS CUBIQUES

Nous appellerons schéma à un $\mathbb{C}$-schéma séparé et de type fini. Les faisceaux sur un schéma X se considèrent relatifs à la topologie classique de X.

On a des résultats analogues aux établis dans ce paragraphe pour les espaces analytiques.

1. Un exemple d'hyperrésolution cubique.

On va donner d'abord un exemple illustratif d'hyperrésolution cubique. Dans la section suivante on donne les définitions formelles qui sont ici seulement suggerées.

Une situation classique qu'on peut considérer du point de vue cubique est la donnée d'une famille finie $\{Y_i\}_{1\leq i\leq n}$ de sous-espaces d'un schéma X. Pour $n=3$ on a un diagramme formé par les morphismes d'inclusion

(1.1)

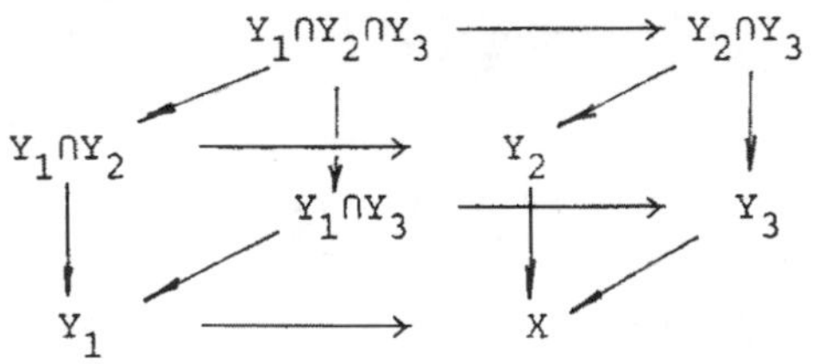

qu'on appelle schéma 3-cubique. On associe à ce diagramme d'espaces le diagramme de complexes de faisceaux

(1.2)

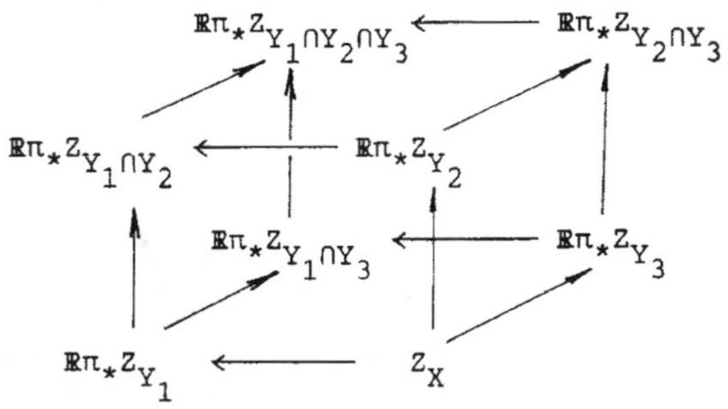

où π dénote les morphismes d'inclusion. Alors on considère ce diagramme comme un complexe multiple. Le complexe double obtenu par contraction de l'indice cubique dans (1.2) est

(1.3) $$Z_X \longrightarrow \bigoplus_{1\le i\le 3} \mathbb{R}\pi_* Z_{Y_i} \longrightarrow \bigoplus_{1\le i<j\le 3} \mathbb{R}\pi_* Z_{Y_i\cap Y_j} \longrightarrow \mathbb{R}\pi_* Z_{Y_1\cap Y_2\cap Y_3} .$$

Si les sous-espaces Y_i forment un recouvrement fermé de X, alors le complexe simple de (1.3) est acyclique, c'est-à-dire le complexe

$$\mathbb{R}\pi_* Z_{Y_\bullet} = s[\oplus\, \mathbb{R}\pi_* Z_{Y_i} \longrightarrow \oplus\, \mathbb{R}\pi_* Z_{Y_i\cap Y_j} \longrightarrow \mathbb{R}\pi_* Z_{Y_1\cap Y_2\cap Y_3}]$$

est quasi-isomorphe à Z_X. On dit dans ce cas que le diagramme (1.1) est de descente cohomologique sur X.

Si en plus les sous-espaces Y_i sont lisses et se coupent transversellement, alors tous les sommets de (1.1) différents de X sont lisses et (1.1) est un exemple d'hyperrésolution cubique du schéma X.

2. Schémas cubiques et leur cohomologie.

(2.1) Soit n un entier positif ou nul. On note $\square_n$ l'ensemble $\{0, 1\}^n$ muni de l'ordre produit. Si $\alpha \in \square_n$ on écrit $|\alpha| = \alpha_1 + \ldots + \alpha_n$. Un schéma n-cubique ou simplement schéma cubique, $X_\bullet = \{X_\alpha\}_{\alpha\in\square_n}$, est la donnée d'un schéma X_α pour chaque $\alpha \in \square_n$ et d'un morphisme $X_{\alpha\beta}\colon X_\beta \to X_\alpha$ pour $\beta \ge \alpha$ tels que $X_{\alpha\gamma} = X_{\alpha\beta} \circ X_{\beta\gamma}$ si $\gamma \ge \beta \ge \alpha$. On note π_α le morphisme $X_{0\alpha}$. Un morphisme $f_\bullet$ d'un schéma n-cubique $X_\bullet$ dans un schéma n-cubique $Y_\bullet$ est la donnée d'un morphisme $f_\alpha\colon X_\alpha \to Y_\alpha$ pour chaque α de façon qu'on ait $Y_{\alpha\beta} \circ f_\beta = f_\alpha \circ X_{\alpha\beta}$.

La définition de morphisme de schémas cubiques peut s'étendre dans le sens suivant. Soit $Y_\bullet$ un schéma m-cubique, et $X_{\bullet\bullet}$ un schéma (m+n)-cubique. On considère le schéma (m+n)-cubique $Y_{\bullet\bullet}$ obtenu à partir de $Y_\bullet$ prenant, pour $\alpha \in \square_m$ et $\beta \in \square_n$,

$$Y_{\alpha\beta} = \begin{cases} Y_\alpha & \text{si } \beta = 0 \\ 0 & \text{si } \beta \neq 0 . \end{cases}$$

Alors un morphisme de $Y_\bullet$ dans $X_{\bullet\bullet}$ est un morphisme de $Y_{\bullet\bullet}$ dans $X_{\bullet\bullet}$. En particulier, si $Y_\bullet$ est la m-face $Y_\alpha = X_{\alpha 0}$, pour $\alpha \in \square_m$, on a un morphisme d'inclusion $Y_\bullet \longrightarrow X_{\bullet\bullet}$.

On dit qu'un schéma n-cubique $X_\cdot$ est propre si tous les morphismes de transition $X_{\alpha\beta}$ sont propres.

Dans ces définitions on peut remplacer $\square_n$ par $\square_n^+ = \square_n - \{0\}$. On obtient alors la notion de schéma cubique diminué que nous désignons par $X_\cdot^+$ ou simplement par $X_\cdot$ s'il n'y a pas craintre de confusion.

(2.2) De façon entièrement analogue au cas des schémas simpliciaux [3] on définit la notion de complexe de faisceaux sur un schéma cubique ainsi que les opérations $f_{\cdot *}$, $f_\cdot^*$ pour un morphisme de schémas cubiques $f_\cdot : X_\cdot \rightarrow Y_\cdot$.

En particulier, si $X_\cdot$ est un schéma cubique et K est un complexe de faisceaux sur $X_\cdot$ on peut considérer le complexe double de faisceaux sur X_0 dont la composante de bidegrée (r, s) est

$$\bigoplus_{|\alpha|=s} \mathbb{R}\pi_{\alpha *} K^{r\alpha} .$$

On désigne par $\mathbb{R}\pi_* K$ le complexe simple associé. On définit la filtration L comme la seconde filtration, associée à l'indice cubique.

(2.3) Soit A un groupe abélien. Si on considère sur chaque sommet X_α d'un schéma cubique $X_\cdot$ le faisceaux constant A_{X_α} , on obtient un faisceau $A_{X_\cdot}$ sur $X_\cdot$. On a une suite exacte de complexes de faisceaux sur X_0 ,

$$0 \longrightarrow A_{X_0} \longrightarrow \mathbb{R}\pi_* A_{X_\cdot^+} \longrightarrow \mathbb{R}\pi_* A_{X_\cdot} [1] \longrightarrow 0 .$$

On définit la cohomologie de $X_\cdot$ à valeurs dans A, $H^*(X_\cdot, A)$, comme l'hypercohomologie du complexe $\mathbb{R}\pi_* A_{X_\cdot}$, c'est à dire,

$$H^n(X_\cdot, A) = \mathbb{H}^n(X_0, \mathbb{R}\pi_* A_{X_\cdot}) .$$

(2.4) Soit $X_\cdot$ un schéma cubique. On dit que $X_\cdot$ est de descente cohomologique si $\mathbb{R}\pi_* Z_{X_\cdot}$ est acyclique ou, de façon équivalente, si le morphisme

$$\mathbb{Z}_{X_0} \to \mathbb{R}\pi_* \mathbb{Z}_{X_\bullet^+}$$

est un quasi-isomorphisme.

(2.5) Si $X_\bullet$ est un schéma cubique de descente cohomologique, la cohomologie de X_0 s'exprime à partir des cohomologies des sommets X_α par la suite spectrale induite par la filtration L,

$$E_1^{pq} = \bigoplus_{|\alpha|=p+1} H^q(X_\alpha, A) \Longrightarrow Gr_L^p H^{p+q}(X_0, A) \;,\; p \geq 0 \;,\; q \geq 0 \;.$$

3. Hyperrésolutions cubiques.

(3.1) La définition d'hyperrésolution cubique qu'on va adopter dans cet exposé est plus faible que l'originale contenue dans [8] mais elle suffit pour les applications à la théorie de Hodge.

Soit $X_\bullet$ un schéma cubique. Nous dirons que $X_\bullet$ est une hyperrésolution cubique de X_0 si $X_\bullet$ est propre, de descente cohomologique et X_α est lisse pour tout $\alpha \neq 0$.

(3.2) Une variante de la définition d'hyperrésolution cubique qui est nécéssaire pour la théorie de Hodge est la suivante.

Soit X un schéma et U un sous-schéma ouvert de X . On appelle hyperrésolution cubique du couple (X, U) une hyperrésolution cubique $X_\bullet$ de X telle que pour chaque $\alpha \neq 0$ l'image inverse Y_α de $Y = X-U$ par $\pi_\alpha : X_\alpha \to X$ soit sur chaque composante irréductible X_α^i de X_α bien un diviseur à croisements normaux de X_α^i , bien X_α^i ou bien vide.

(3.3) Nous allons décrire la récurrence qui fournit la construction d'une hyperrésolution cubique d'un schéma X .

(3.3.1) Le premier pas de la récurrence est la résolution des singularités de X . Puisque X est, en général, réductible il faut considérer ses composantes irréductibles $\{X^i\}_{i\in I}$. Soit, pour chaque $i \in I$, $\pi^i : \tilde{X}^i \to X^i$ une résolution des singularités de X^i , et désignons par $\pi : \tilde{X} \to X$ la somme disjointe des π^i , $i \in I$. Alors π est un morphisme propre qui est un isomorphisme hors d'une sous-variété fermée minimal $\Delta = \Delta(\pi)$ de X , de dimension plus petite que la dimension de X. Si Y est un fermé de X qui contient Δ , d'après le théorème de changement de base pour les morphismes propres, le carré cartésien

$$\begin{array}{ccc} \tilde{Y} & \longrightarrow & \tilde{X} \\ \downarrow & & \downarrow \pi \\ Y & \longrightarrow & X \end{array}$$

est un schéma 2-cubique de descente cohomologique.

(3.3.2) Posons $Y = \Delta$ et désignons le carré résultant par

$$\begin{array}{ccc} X_{11} & \longrightarrow & X_{10} \\ \downarrow & & \downarrow \\ X_{01} & \longrightarrow & X_{00} \end{array} .$$

Si X_{01} et X_{11} sont lisses ce diagramme est une hyperrésolution cubique de X. Dans le cas contraire on doit suivre la récurrence et remplacer les schémas X_{01} et X_{11} par des diagrammes de variétés lisses ou de dimension inférieure.

(3.3.3) On considère maintenant des résolutions de X_{01} et X_{11}. Néanmoins pour que les diagrammes du type (3.3.1) qu'en résultent puissent être ensemblés de façon à donner un schéma 3-cubique, on procède de la façon suivante.

Soient $\{X^i_{11}\}_{i \in I_1}$ les composantes irréductibles de X_{11} et $X^i_{01} = \mathrm{Image}(\pi\colon X^i_{11} \to X_{00})$, pour $i \in I_1$. On considère, pour chaque $i \in I_1$, une résolution de singularités, $\tilde{X}^i_{01} \to X^i_{01}$, et une résolution de singularités de l'adhérence, dans le produit fibré de $\tilde{X}^i_{01}$ par X^i_{11} au dessus de X^i_{01}, de l'ouvert dense de X^i_{01} complémentaire de $\Delta(\tilde{X}^i_{01} \to X^i_{01})$. Si on définit $\tilde{X}_{j1} \to X_{j1}$ comme la somme disjointe de la famille $\{\tilde{X}^i_{j1} \to X^i_{j1}\}_{i \in I_1}$, pour $j = 0, 1$, on obtient un carré commutatif

$$\begin{array}{ccc} \tilde{X}_{11} & \longrightarrow & X_{11} \\ \downarrow & & \downarrow \\ \tilde{X}_{01} & \longrightarrow & X_{01} \end{array} .$$

Posons maintenant $Y_{11} = \Delta(\widetilde{X}_{11} \to X_{11})$, et $Y_{01} = \mathrm{Image}(Y_{11} \to X_{01})$, et définons $\widetilde{Y}_{j1}$ comme le produit fibré de Y_{j1} par $\widetilde{X}_{j1}$ au dessus de X_{j1} , pour $j = 0,1$. On obtient un diagramme commutatif

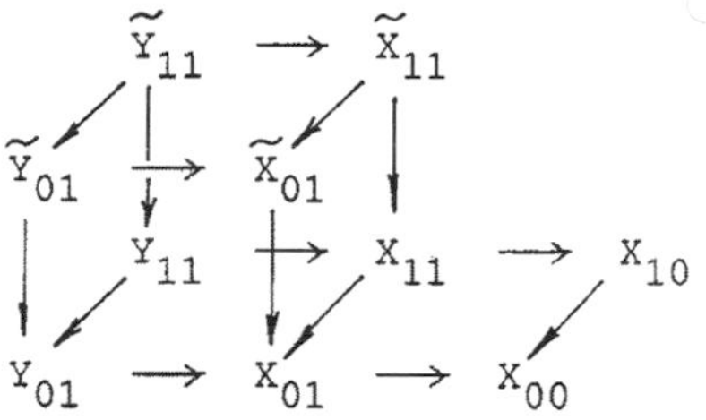

qu'on écrit

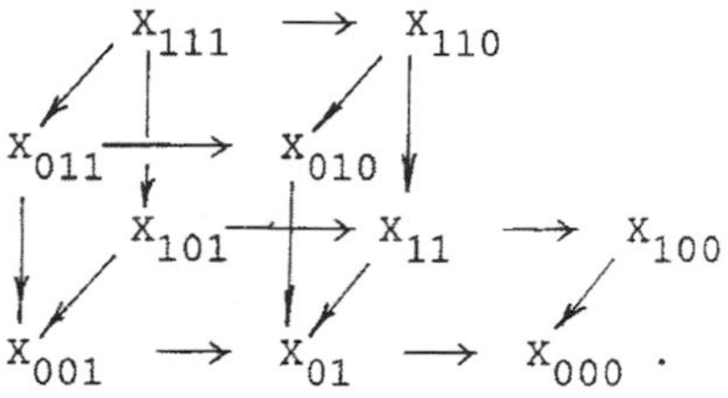

X_{111}
X_{110}
X_{011}
X_{010}
X_{101}
X_{11}
X_{100}
X_{001}
X_{01}
X_{000} .

D'après le théorème de changement de base appliqué aux résolutions de X_{01} et X_{11} il résulte que le schéma 3-cubique propre définit par les X_α , $\alpha \in \square_3$, est de descente cohomologique.

On a, pour $\alpha \in \square_3^+$, $\dim X_\alpha \leq \dim X - |\alpha| + 1$ et X_α est lisse si $\alpha_3 = 0$. Si $\dim X = 2$, les X_α , avec $\alpha_3 = 1$, ont dimension nulle, ils sont donc lisses et $\{X_\alpha\}_{\alpha \in \square_3}$ est une hyperrésolution de X . En general il faut résoudre la face singulière $\{X_\alpha\}_{\alpha \in \square_3}$, $\alpha_3 = 1$.

(3.3.4) L'étape n-ième de la récurrence est analogue, il faut cependant remarquer qu'un point important pour construire une résolution de la face singulière $\{X_\alpha\}_{\alpha \in \square_n}$, $\alpha_n = 1$, est la décomposition d'un schéma cubique en composantes irréductibles, dont la définition est la suivante.

Soit $X_\bullet$ un schéma cubique. Un sous-espace fermé de $X_\bullet$ est un schéma cubique $Y_\bullet$ muni d'un morphisme de schémas cubiques $Y_\bullet \to X_\bullet$ qui est une immersion fermée dans chaque sommet. On dit que $X_\bullet$ est irréductible s'il n'est pas vide et si la réunion de deux sous-espaces fermés de $X_{\bullet}$ différents de $X_{\bullet}$ est toujours différente de $X_{\bullet}$. Une composante irréductible de $X_{\bullet}$ est un sous-espace irréductible maximal de $X_{\bullet}$. Il résulte aussitôt que $X_{\bullet}$ est la réunion de ses composantes irréductibles.

(3.3.5) D'après l'étape n-ième on obtient un schéma (n+1)-cubique $\{X_\alpha\}_{\alpha \in \square_{n+1}}$, qui est propre, de descente cohomologique et tel que, pour $\alpha \in \square^{+}_{n+1}$, $\dim X_\alpha \leq \dim X - |\alpha| + 1$ et X_α est lisse si $\alpha_{n+1} = 0$.

Si $n = \dim X$ les X_α , avec $\alpha_{n+1} = 1$, ont dimension nulle, ils sont donc lisses et $\{X_\alpha\}_{\alpha \in \square_{n+1}}$ est une hyperrésolution de X .

Ceci démontre le théorème principal:

(3.4) <u>Théorème</u> [8]. Si X est un schéma de dimension N il existe un schéma (N+1)-cubique $X_{\bullet}$ qui est une hyperrésolution cubique de $X_0 = X$ et qui satisfait

$$\dim X_\alpha \leq \dim X - |\alpha| + 1$$

pour tout $\alpha \in \square^{+}_{N+1}$.

(3.5) Dans les applications des hyperrésolutions cubiques à la théorie de Hodge-Deligne on a besoin de comparer deux hyperrésolutions cubiques d'un même schéma.

Pour les résolutions ordinaires, si on considère deux résolutions des singularités d'un schéma X , $\tilde{X}' \to X$ et $\tilde{X}'' \to X$, il existe une troisième résolution $\tilde{X} \to X$ et des morphismes convenables qui rendent commutatif le diagramme

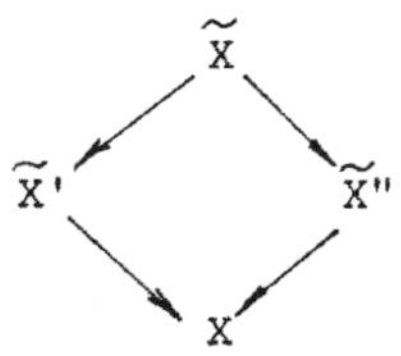

Pour les hyperrésolutions il y a un résultat plus faible mais suffisant pour les applications:

<u>Théorème</u> [8]. La catégorie des hyperrésolutions cubiques d'un schéma X est connexe. Plus précisement, données deux hyperrésolutions cubi-

ques $X'_\bullet$ et $X''_\bullet$ de X il existe des hyperrésolutions cubiques $\widetilde{X}_\bullet$, $\widetilde{X}'_\bullet$ et $\widetilde{X}''_\bullet$ et un diagramme de morphismes de schémas cubiques

$$X'_\bullet \longrightarrow \widetilde{X}'_\bullet \longleftarrow \widetilde{X}_\bullet \longrightarrow \widetilde{X}''_\bullet \longleftarrow X''_\bullet$$

qui sont l'identité sur X .

En effet, on remarque d'abord que la notion et la construction d'hyperrésolution cubique d'un schéma peut s'étendre à certains types de diagrammes de schémas. En particulier on peut considérer des hyperrésolutions cubiques de schémas cubiques. Un autre type de diagramme qui admet des hyperrésolutions cubiques est obtenu à partir de deux schémas cubiques $X'_\bullet$ et $X''_\bullet$ avec un même sommet final, en identifiant ce sommet. Si $X'_\bullet$ et $X''_\bullet$ sont deux hyperrésolutions cubiques du schéma X , on a $X'_0 = X = X''_0$. Considérons le diagramme obtenue en identifiant X'_0 avec X''_0 , qui nous écrirons d'une façon abrégée

$$X'_\bullet \longleftarrow X \longrightarrow X''_\bullet \ .$$

Il existe une hyperrésolution cubique de ce diagramme,

$$\widetilde{X}'_{\bullet\bullet} \longleftarrow \widetilde{X}_\bullet \longrightarrow \widetilde{X}''_{\bullet\bullet}$$

où $\widetilde{X}'_{\bullet\bullet}$ (resp. $\widetilde{X}''_{\bullet\bullet}$) est un diagramme $(n'+n)$-cubique (resp. $(n''+n)$-cubique) qui est une hyperrésolution cubique de $X'_\bullet$ (resp. $X''_\bullet$) et tel que $\widetilde{X}'_{0\bullet} = \widetilde{X}_\bullet$ (resp. $\widetilde{X}''_{0\bullet} = \widetilde{X}_\bullet$) est une hyperrésolution cubique de X . Alors $\widetilde{X}'_{\bullet\bullet}$ et $\widetilde{X}''_{\bullet\bullet}$ sont des hyperrésolutions cubiques de X et on a des morphismes d'inclusion

$$X'_\bullet \longrightarrow \widetilde{X}'_{\bullet\bullet} \longleftarrow \widetilde{X}_\bullet \longrightarrow \widetilde{X}''_{\bullet\bullet} \longleftarrow X''_\bullet$$

ce qui prouve le théorème.

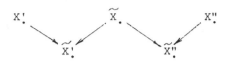

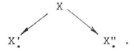

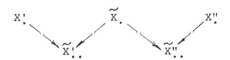

(3.6) On a des résultats analogues à (3.4) et (3.5) pour le cas d'un couple (X, U), (3.2).

§ II. HYPERRESOLUTIONS CUBIQUES ET THEORIE DE HODGE-DELIGNE

Comme nous avons rappellé dans l'introduction, la méthode introduite par Deligne pour munir d'une structure de Hodge mixte les groupes de cohomologie d'un schéma X consiste à remplacer X par un schéma simplicial qui est le complementaire d'un diviseur à croisements normaux dans un schéma simplicial propre et lisse. Ceci permet d'obtenir un complexe de Hodge mixte qui munit d'une structure de Hodge mixte la cohomologie de X. Dans cette contexte les hyperrésolutions cubiques fournissent un instrument alternatif aux hyperrecouvrements simpliciaux propres et lisses de Deligne pour obtenir un complexe de Hodge mixte qui induit sur la cohomologie de X la même structure de Hodge mixte.

1. Hyperrésolutions cubiques et structures de Hodge mixtes.

(1.1) Soit X un schéma, $\bar{X}$ une compactification de X et $\bar{X}_\bullet$ une hyperrésolution cubique du couple $(\bar{X}, X)$. On a donc une situation analogue à [3] (8.8.19) pour obtenir un complexe de Hodge mixte cohomologique $K_{\bar{X}}$ sur $\bar{X}$ dont l'hypercohomologie coïncide avec la cohomologie de X. La proposition qui suit donne des précisions sur la filtration par le poids de $K_{\bar{X}}$. Elle résulte de [3] (7.1.6) et (3.1.5.2) appliqués à une hyperrésolution cubique $\bar{X}_\bullet$ de $(\bar{X}, X)$ telle que $\dim \bar{X}_\alpha \leq N-|\alpha| = 1$.

(1.2) Proposition. Soit X un schéma de dimension N et $\bar{X}$ une compactification de X. Alors il existe un complexe de Hodge mixte cohomologique $K_{\bar{X}}$ sur $\bar{X}$, tel que $\mathbb{H}^*(\bar{X}, K_{\bar{X}}) \cong H(X, Z)$ et qui induit sur les groupes de cohomologie de X la structure de Hodge mixte de Deligne. En plus $K_{\bar{X}}$ satisfait

i) la filtration par le poids W de $K_{\bar{X}}$ vérifie que $W_{-N-1}=0$ et $W_N=K_{\bar{X}}$,

ii) si les faisceaux de cohomologie $h^n(Gr_q^W K_{\bar{X}})$ sont non nuls on a $q \leq n \leq 2N+q$, $0 \leq n+q \leq 2N$,

iii) la dimension du support des faisceaux $h^n(Gr_q^W K_{\bar{X}})$ est $\leq N-n$.

(1.3) Remarque. De la suite spectrale

$$E_2^{p,q} = H^p(X, h^q(Gr_r^W K)) \Longrightarrow H^{p+q}(X, Gr_r^W K)$$

et de (1.2) on obtient les renseignements obtenus par Deligne [3] (8.2.4) pour la filtration par le poids sur la cohomologie de X .

(1.4) Pour un schéma cubique $X_.$, on a un résultat analogue à (1.2).

(1.5) La structure de Hodge mixte des groupes de cohomologie relative a été obtenue par Deligne [3] (8.3.8). Ici nous remarquons le fait que du point de vue cubique le cas d'un morphisme se ramène naturellement au cas dejà consideré d'un schéma cubique. En effet, si $f: X_. \longrightarrow Y_.$ est un morphisme de schémas cubiques, on associe à $f_.$ le schéma cubique $Z_{..}$ tel que $Z_{0.} = Y_.$, $Z_{1.} = X_.$ et la structure de Hodge mixte du morphisme est, par définition, celle de $Z_{..}$. Ceci évite le passage au schéma simplicial $C(f)$, [3] (6.3).

En particulier, si X est un schéma et $\bar{X}$ est une compactification de X on a l'isomorphisme

(1.5.1) $$H_c^n(X) \cong H^n(\bar{X}-X \longrightarrow X)$$

à travers lequel on munit les groupes de cohomologie à support compact d'une structure de Hodge mixte fonctorielle pour les morphismes propres.

De même si Y est un sous-schéma fermé de X on a l'isomorphisme

(1.5.2) $$H_Y^n(X) \cong H^n(X-Y \longrightarrow X)$$

à travers lequel on munit les groupes de cohomologie locale d'une structure de Hodge mixte fonctorielle en $(X, X-Y)$.

(1.6) Si $X_.$ est un schéma cubique on peut considérer la suite spectrale associée a l'indice cubique. Elle est une suite spectrale de structures de Hodge mixtes. Plus généralement si on considère la décomposition $\square_{m+n} = \square_m \times \square_n$, et la filtration induite par l'indice cubique de $\square_m$ on obtient le résultat suivant.

Proposition. Soit $X_{..}$ un $(m+n)$-schéma cubique. On a la suite spectrale de structures de Hodge mixtes suivante

$$E_1^{pq} = \bigoplus_{\alpha} H^q(X_{\alpha\cdot}, Z) \Longrightarrow H^{p+q}(X_{\cdot\cdot}, Z)$$

où $|\alpha| = p$, $\alpha \in \square_m$.

(1.7) De la proposition antérieure et de (1.6.1), (1.6.2) on obtient en particulier que les suites exactes de cohomologie à support compact, de cohomologie locale, et de Mayer-Vietoris sont suites exactes de structures de Hodge mixtes.

Nous allons expliciter un exemple.

(1.7.1) Soit $f: X' \longrightarrow X$ un morphisme propre de schémas et Y un sous-espace fermé de X tel que f est un isomorphisme en dehors de Y ; soit $Y' = f^{-1}(Y)$, Z un fermé de X et $Z' = f^{-1}(Z)$. Alors la suite exacte de cohomologie entière

$$\cdots \longrightarrow H_Z^q(X) \longrightarrow H_{Z'}^q(X') \oplus H_{Y\cap Z}^q(Y) \longrightarrow H_{Y'\cap Z'}^q(Y') \longrightarrow H_Z^{q+1}(X) \longrightarrow \cdots$$

est une suite exacte de structures de Hodge mixtes.

En effet, on considère le schéma cubique $X_{\cdot\cdot}$ défini par

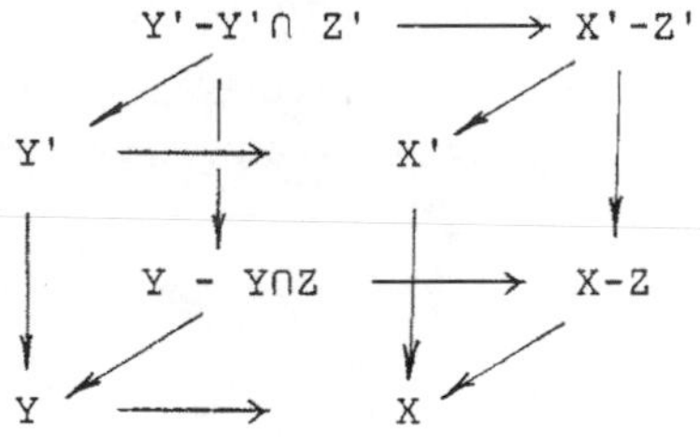

où les morphismes sont induits par f ou bien par des inclusions. La suite exacte antérieure résulte alors de (1.6) et du fait que $X_{\cdot\cdot}$ est de descente cohomologique.

2. Structures de Hodge mixtes associées aux espaces analytiques.

(2.1) Si X est un espace analytique réduit et Y un sous-espace qui est une variété algébrique compacte les groupes de cohomologie locale $H_Y^*(X, Z)$ peuvent se munir d'une structure de Hodge mixte qui ne dépend que du germe. D'après la méthode décrite, on considère une hyperrésolution cubique $X_\cdot$ de $(X, X-Y)$ et on se ramène à obtenir d'une façon fonctorielle un complex de Hodge mixte cohomologique (pour abréger CHMC) convenable sur chaque X_α, $\alpha \neq 0$. C'est à dire, il faut considérer d'abord le cas où X est lisse et Y est un diviseur à croisements normaux dans X qui est une variété algébrique compacte. Alors on remarque que dans ce cas les groupes $H^*(X, \mathbb{Z})$ ne possèdent pas, en général, une structure de Hodge mixte et par conséquant on ne peut pas appliquer la méthode correspondante au cas algébrique. Cette difficulté se soulève en utilisant la variante suivante introduite par Fujiki [6] (cf.[5]) et qui dans le cas où X soit algébrique amène à la structure de Hodge mixte ordinaire de $H_Y^*(X, \mathbb{Z})$.

Soit $X^* = X - Y$ et j l'inclusion de X^* dans X. On considère les complexes

$$K_{\mathbb{Z}} = \mathbb{R}\Gamma_Y \, \mathbb{Z}_X$$
$$K_{\mathbb{Q}} = s(\mathbb{Q}_X \to \mathbb{R}j_* \, \mathbb{Q}_{X^*})$$
$$K_{\mathbb{C}} = s(\Omega_X^{\bullet} \to \Omega_X^{\bullet}(\log Y))$$

avec les filtrations

$$W\,K_{\mathbb{Q}} = \tau[-2] \text{ (filtration canonique decalée)}$$
$$W_r K_{\mathbb{C}} = W_{r+1}\, \Omega_X^{\bullet} \oplus (W_{r+1}\, \Omega_X^{\bullet}(\log Y)[-1])$$
$$F^p K_{\mathbb{C}} = F^p\, \Omega_X^{\bullet} \oplus (F^p\, \Omega_X^{\bullet}(\log Y)[-1])$$

Alors on a que

(2.1.1) $K_{X,Y} = (K_{\mathbb{Z}}, (K_{\mathbb{Q}}, W), (K_{\mathbb{C}}, W, F))$

est un CHMC qui munit les groupes $H_Y^*(X, \mathbb{Z})$ d'une structure de Hodge mixte fonctorielle en (X, X^*).

(2.2) Sous les hypothèses générales de (2.1) on suppose que l'espace analytique X se rétracte sur Y. Alors $H^*(Y, \mathbb{Z}) \cong H^*(X, \mathbb{Z})$ et par conséquent les groupes $H^*(X, \mathbb{Z})$ sont munis d'une structure de Hodge mixte induite par l'isomorphisme antérieur. Dans ce cas il est possible aussi de définir une structure de Hodge mixte sur le noeud $X^* = X - Y$. L'idée est, naturellement, de remplacer l'espace X^* par le diagramme

$$\begin{array}{ccc} \emptyset & \longrightarrow & Y \\ \downarrow & & \downarrow \\ X^* & \longrightarrow & X \end{array} \qquad (2.2.1)$$

dont la cohomologie est celle de X^* .

Comme avant on se ramène au cas où X est lisse et Y est un diviseur à croisements normaux dans X qui est une variété algébrique compacte, telle que X se retracte sur Y. Alors dans le diagramme ci-dessus les groupes de cohomologie de $X^* \to X$ et Y possèdent une structure de Hodge mixte. L'incarnation de ces structures de Hodge mixtes au niveau de CHMC permet d'obtenir un CHMC sur X, K_{X^*}, qu'induit une structure de Hodge mixte sur $H^*(X^*, Z)$. Plus précisement, le CHMC qui correspond à $X^* \to X$ a été décrit dans (2.1). Quant à Y on doit prendre un CHMC qui soit fonctoriel en Y . Par exemple on peut prendre le CHMC, K_Y , définit par le schéma simplicial associé a la subdivision baricentrique du recouvrement $Y = \cup Y_i$ par les composantes irréductibles de Y [12] (12.1). Alors K_{X^*}est le simple de Deligne [3](3.7.1.6) du morphisme, $K_{X,Y} \to K_Y$, induit par le diagramme (2.2.1).

(2.3) En définitive on a le théorème suivant

<u>Théorème</u> [8]. Soit X un espace analytique réduit et Y un sous-espace qui est une variété algébrique compacte. Alors

i) les groupes de cohomologie locale $H_Y^*(X, Z)$ ont une structure de Hodge mixte fonctorielle en (X, X^*) qui ne dépend que du germe,

ii) si X se retracte sur Y , les groupes $H^*(X^*, Z)$ sont munis d'une structure de Hodge mixte fonctorielle qui ne dépend que du germe,

iii) la suite exacte de cohomologie locale

$$\ldots \to H_Y^i(X, Z) \to H^i(X, Z) \to H^i(X^*, Z) \to H_Y^{i+1}(X, Z) \to \ldots$$

est une suite exacte de structures de Hodge mixtes.

(2.3.1) Il existe des contrôles locaux et globaux sur la filtration par le poids de ces structures de Hodge mixtes qui s'obtiennent d'une façon analogue au cas algébrique, (1.2) et (1.3).

§ III. STRUCTURE DE HODGE MIXTE DANS L'HOMOLOGIE ET FILTRATION DE ZEEMAN.

Dans cette section on donne d'après [7] une application des résultats antérieurs pour montrer une relation entre la filtration de Zeeman et la filtration par le poids, qui avait été conjecturée par McCrory [11], et demontrée par lui même pour les hypersurfaces [11].

1. Rappels sur la dualité.

Pour ce qui suit nous nous réferons à [1] et [14].

(1.1) Soit X un schéma et k un anneau commutatif de Gorenstein (nous considérons uniquement les cas $\mathbb{Z}$, $\mathbb{Q}$ ou $\mathbb{C}$), il existe un complexe de faisceaux de k-modules sur X denoté par $\mathbb{D}_X(k)$, dont l'hypercohomologie est l'homologie de Borel-Moore de X (c'est à dire, l'homologie du complexe de chaînes infinies localement finies ou sans condition sur le support) à coefficients dans k,

$$H_i(X, k) = \mathbb{H}^{-i}(X, \mathbb{D}_X(k)) .$$

Si $K^{\bullet}$ est un complexe de faisceaux de k-modules sur X on défini $D(K^{\bullet}) = \mathbb{R}\mathrm{Hom}_k^{\bullet}(K^{\bullet}, \mathbb{D}_X)$. Le foncteur D est dualisant dans le sens suivant: pour tout morphisme de schémas $f: X \to Y$ et tout complexe de faisceaux de k-modules sur X à cohomologie bornée, $K^{\bullet}$, on a un quasi-isomorphisme

$$\mathbb{R}f_* \, \mathbb{R}\, \underline{\mathrm{Hom}}_k^{\bullet}(K^{\bullet}, \mathbb{D}_X(k)) \cong \mathbb{R}\, \underline{\mathrm{Hom}}_k^{\bullet}(\mathbb{R}f_!\, K^{\bullet}, \mathbb{D}_Y(k)) .$$

En particulier, si k est un corps, $H_i(X)$ est dual de $H_c^i(X)$.

(1.2) Nous aurons besoin d'expliciter le foncteur dualisant $D(K^{\bullet})$ dans le cas particulier où k est un corps et $K^{\bullet}$ un complexe borné de faisceaux c-mous. Avec ces hypothèses $D(K^{\bullet})$ est incarné par le faisceaux dont les sections sur un ouvert U de X sont

$$\mathrm{Hom}_k(\Gamma_c(U, K^{\bullet}), k) .$$

En outre, si $K^\bullet$ est muni d'une filtration $F^\bullet$ on définit la filtration $D(F^\bullet)$, duale de $F^\bullet$, sur $D(K^\bullet)$ par

$$D(F)_p\, D(K) = D(K^\bullet/F^{p+1}K^\bullet) .$$

Si les sous-complexes $F^pK^\bullet$ sont c-mous, alors $D(F)_p$ est défini sur un ouvert U de X par

$$\mathrm{Hom}_k(\Gamma_c(U, K^\bullet/F^{p+1}K^\bullet), k) .$$

(1.3) Considérons un schéma lisse X et le complexe double de faisceaux $\mathcal{E}_X^{\bullet\bullet}$ des germes de formes différentielles C^∞ à valeurs complexes, muni de la filtration F de Hodge. Le complexe dual est formé des "courants algébriques" (voir [6]). C'est le complexe double défini par

$$\mathcal{D}_X^{p,q} = \mathrm{Hom}_{\mathbb{C}}(\Gamma_c(U, \mathcal{E}_X^{-p,-q}), \mathbb{C}) .$$

On ne considère aucune topologie sur $\Gamma_c(U, \mathcal{E}_X^{\bullet\bullet})$.
Compte tenu que $\mathcal{E}_X^{\bullet\bullet}$ est une résolution du faisceau $\mathbb{C}_X$, le dual $\mathcal{D}_X^{\bullet\bullet}$ est une incarnation du complexe dualisant à coefficients complexes. Ce complexe $\mathcal{D}_X^{\bullet\bullet}$ est muni de la filtration de Hodge $D(F^\bullet)$ dénotée aussi par $F^\bullet$.

Il résulte aussitôt que si X est compact, alors $(\mathbb{D}_X(\mathbb{Z}), (\mathcal{D}_X^{\bullet\bullet}, F^\bullet))$ est un CHC sur X qui induit sur $H_*(X)$ la structure de Hodge mixte duale à celle de $H^*(X)$.

La construction antérieure est fonctorielle pour les morphismes propres: si $\pi: X_1 \to X_2$ est un morphisme propre de schémas lisses, alors on a un morphisme filtré

$$\pi_*: (\pi_*\mathcal{D}_{X_1}^{\bullet\bullet}, \pi_*F) \longrightarrow (\mathcal{D}_{X_2}^{\bullet\bullet}, F)$$

qui est dual du morphisme filtré

$$\pi^*: (\mathcal{E}_{X_2}^{\bullet\bullet}, F) \longrightarrow (\pi_*\mathcal{E}_{X_1}^{\bullet\bullet}, \pi_*F) .$$

2. Structure de Hodge mixte sur l'homologie.

(2.1) Le groupe d'homologie $H_i(X)$ d'un schéma X est dual du groupe $H_c^i(X)$, donc $H_i(X)$ est munie d'une structure de Hodge mixte, duale à celle de $H^i_c(X)$.

Il s'agit ici d'obtenir cette structure à partir d'un CHMC. Pour simplifier les notations nous considérons uniquement la partie définie sur $\mathbb{C}$.

(2.2) Si X est un schéma compact et $X_\bullet$ est une hyperrésolution cubique de $X_\bullet$, compte tenu de (1.3), il résulte que le dual de $(\pi_* \mathcal{E}^{\bullet\bullet}_{X^+_\bullet}, L, F)$ est un CHMC sur X ,

$$(\pi_* \mathcal{D}^{\bullet\bullet}_{X^+_\bullet}, L, F) ,$$

qui induit sur $H_*(X)$ la structure de Hodge mixte duale à celle de $H^*(X)$.

(2.3) Si X est un schéma arbitraire, il faut considérer une compactification $X \to \bar{X}$ de X et une hyperrésolution cubique du couple $(\bar{X}, X)$. Le problème révient alors à considèrer pour un schéma lisse X et une compactification $j: X \to \bar{X}$ de X tel que $\bar{X} - X$ soit un diviseur à croisements normaux de $\bar{X}$, un CHMC sur $\bar{X}$ dont le complexe sous-jacent soit $\mathbb{R}j_*\mathbb{D}_X$, et qui soit fonctorielle en $(\bar{X}, X)$.
Il suffit de prendre le cône du dual du morphisme

$$\mathcal{E}^{\bullet\bullet}_X \longrightarrow \mathcal{E}^{\bullet\bullet}_{N(Y)_\bullet}$$

où $N(Y)_\bullet$ est le schéma simplicial associé à la subdivision baricentrique du recouvrement $Y = \cup Y_i$ des composantes irréductibles de Y [12](12.1), la filtration par le poids étant la filtration L .

3. Preuve de la conjecture de McCrory.

Soit X un schéma compact de dimension N .

(3.1) La filtration de Zeeman S sur l'homologie $H_i(X)$ est une filtration topologique, définie sur tout anneau de coefficients et telle que $S^0H_i(X) = H_i(X)$ et $S^{2N-i+1}H_i(X) = 0$.

Cette filtration avait étée etudiée par première fois par Zeeman dans sa thèse [15]. Après McCrory [10] a donné une intéressante caractérisation en termes d'une stratification: si on a une stratification de Whitney $X = X_N \supset X_{N-1} \supset \ldots \supset X_0$ du schéma X , alors $z \in S^qH_i(X)$ si et seulement si z est représenté par un PL-cycle Z tel que

$|Z| \cap X_k$ ait codimension $\geq q$ dans X_k pour tout k .

(3.2) Du point vue local la filtration de Zeeman peut être définie sur le complexe dualisant $\mathbb{D}_X$ et on a la caractérisation suivante: si τ dénote la filtration canonique de $\mathbb{D}_X$, alors $\tau = \text{Dec } S$. Ceci entraîne que sur $H_i(X)$ on a la relation

$$\tau_q \, H_i(X) = S^{-q-i} \, H_i(X) \, .$$

(3.3) Soit $X_\bullet$ une hyperrésolution cubique de X , telle que $\dim X_\alpha \leq N - |\alpha| + 1$, si $\alpha \neq 0$, alors le complexe double

$$D^{ij} = \bigoplus_{|\alpha|=-j+1} \pi_* \mathcal{D}^i_{X_\alpha} \, , \quad j \leq 0$$

vérifie que $D^{ij} = 0$ si $i < 2N-2j$.

On a sur $sD^{\cdot\cdot}$ la filtration

$$L_q \, sD^{\cdot\cdot} = \bigoplus_{j \geq -q} D^{ij}$$

qui induit une filtration, denotée aussi par L , sur l'homologie $H_i(X, \mathbb{C})$ qui vérifie

$$L_q \, H_i(X, \mathbb{C}) = W^{-q+i} \, H_i(X, \mathbb{C})$$

où W est la filtration par le poids de Deligne.

Étant donné que la filtration τ vérifie

$$\tau_q \, sD^{\cdot\cdot} \subset \bigoplus_{i+j \leq q} D^{ij}$$

il résulte que $\tau_{q-2N} \, sD^{\cdot\cdot} \subset L_q \, sD^{\cdot\cdot}$. Puisque les filtrations S et W sont définies sur $\mathbb{Q}$ on obtient le théorème suivant.

Théorème [7]. Soit X un schéma compact, S et W les filtrations de Zeeman et par le poids de Deligne, respectivement, sur l'homologie rationnelle de X , alors on a la relation

$$S^{2N-i-q} H_i(X, \mathbb{Q}) \subset W^{i-q} H_i(X, \mathbb{Q})$$

pour tout i, q .

§ IV. STRUCTURE DE HODGE MIXTE LIMITE DANS LE CAS D'UNE FAMILLE DE VARIETES ALGEBRIQUES COMPACTES

1. Rappels sur les résultats de Steenbrink.

(1.1) Soit D le disque $\{t \in \mathbb{C} \ ; \ |t|<\eta\}$ où le rayon $\eta > 0$ est choisi convenablement petit selon le contexte, $D^* = D - \{0\}$, $p: \tilde{D}^* \to D^*$ le revêtement universel de D^* défini par $\tilde{D}^* = \{u \in \mathbb{C} \ ; \ \mathrm{Im}\, u > 0\}$ et $p(u) = \eta \exp(2\pi i u)$.

Soit X un espace analytique et $f: X \to D$ un morphisme algébrique et propre. On définit la fibre limite par $\tilde{X}^* = X \times_D \tilde{D}^*$ et on écrit $X^* = X \times_D D^*$, $Y = f^{-1}(0)$, $X_t = f^{-1}(t)$, $t \neq 0$. On dénote par $j: X^* \to X$, $i: Y \to X$, $k: \tilde{X}^* \to X$ les morphismes d'inclusion et projection respectivement.

On a donc les diagrammes cartésiens

$$\begin{array}{ccccc} \tilde{X}^* & \xrightarrow{k} & X & \xleftarrow{i} & Y \\ \downarrow & \searrow \; X^* \nearrow_{j} & \downarrow f & & \downarrow \\ \tilde{D}^* & \xrightarrow[p]{} & D & \longleftarrow & 0 \end{array}$$

Si η est choisi convenablement petit, $f: X \to D$ induit une fibration topologique localemente triviale $f^*: X - Y \to D^*$, ce qu'on supposera dans toute la suite.

Comme D^* est contractile on a l'isomorphisme $H^n(\tilde{X}^*) \cong H^n(X_t)$, $t \neq 0$ et si on considère le complexe de faisceaux défini par $\psi\mathbb{C}_X = i^{-1}\,\mathbb{R}k_*\,\mathbb{C}_{\tilde{X}^*}$ on a un isomorphisme naturel $H^n(\tilde{X}^*, \mathbb{C}) \cong \mathbb{H}(Y, \psi\mathbb{C}_X)$.

(1.2) Le groupe fondamental $\pi_*(D^*)$ est isomorphe via l'application p au groupe de transformations de $\tilde{D}^*$ donné par $\{u \to u+m \ , \ m \in \mathbb{Z}\}$. Soit γ un générateur positif de $\pi_1(D^*)$ et T_t l'isomorphisme de monodromie de $H^n(X_t)$, $t \neq 0$, induit par γ. Soit maintenant T l'action sur $k_*\mathbb{C}_{\tilde{X}^*}$ induite par le générateur $u \to u+1$. Cette action se manifeste par fonctorialité sur $\psi\mathbb{C}_X$ et sur $H^n(\tilde{X}^*)$ pour tout n. On la dénote aussi par T et on voit que T et T_t sont compatibles avec l'isomorphisme $H^n(\tilde{X}^*) \cong H^n(X_t)$.

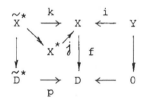

L'endomorphisme T est quasiunipotent et si ρ est l'index de quasiunipotence, i.e. ρ est le plus petit entier positif tel que $(T^{\rho} - I)^m = 0$. On définit $N = \log T^{\rho}$ qui est un endomorphisme nilpotent de $H^n(X^*)$. Si $\rho = 1$ on dit que T est unipotent.

On note $H^n(\tilde{X}^*)_u$ le sous-espace maximal de $H^n(\tilde{X}^*)$ où T opère de façon unipotente; alors T_u est la restriction de T a $H^n(\tilde{X}^*)_u$ et $N_u = \log T_u$.

(1.3) Suivant une démarche qui est analogue à la décrite au § II il faut commencer pour considérer le cas $f\colon X \to D$ où X est une variété analytique (lisse) connexe, f est holomorphe, plat et propre, $Y = f^{-1}(0)$ est un diviseur à croisements normaux dans X et f est lisse sur $X^* = X-Y$. Nous allons exposer les résultats de Steenbrink dont nous aurons besoin. Ils sont contenus dans [12], où nous renvoyons pour les notations et les preuves.

Pour $\alpha \in \mathbb{Q}$, $0 \le \alpha < 1$, $L^{\bullet}_{\alpha}$ est le complexe de faisceaux sur Y dont les sections locales sont de la forme

$$t^{-\alpha} \sum_{i=0}^{s} \theta_i (\log t)^i$$

avec $\theta_0, \ldots, \theta_s$ sections de $i^{-1}\, \Omega^{\bullet}_X(\log Y)$. $L^{\bullet}_{\alpha}$ est un sous-complexe de $i^{-1}k_*\, \Omega^{\bullet}_{X^*}$ tel que T opère de façon unipotente pour $\alpha=0$. Soit $L = \oplus L^{\bullet}_{\alpha}$. Alors, compte tenue la remarque ci-dessus, que la différentielle d du complexe L est telle que $dL_{\alpha} \subset L_{\alpha}$, et la suite spectrale

$$E_2^{pq} = H^p(Y, h^q(L_{\alpha})) \Longrightarrow \mathbb{H}^n(Y, L_{\alpha})$$

on obtient $H^n(\tilde{X}^*)_u \cong \mathbb{H}^n(Y, L_0)$. Donc le complexe de Hodge mixte cohomologique sur Y introduït par Steenbrink

$$A^{\bullet}_X = (A^{\bullet}_Z, (A^{\bullet}_{\mathbb{Q}}, W), (A^{\bullet}_{\mathbb{C}}, W, F))$$

est tel que

$$H^*(\tilde{X}^*)_u \cong \mathbb{H}^n(Y, A^{\bullet}_{\mathbb{C}}) ,$$

c'est à dire $A^{\bullet}_X$ munit d'une structure de Hodge mixte la partie unipotente des groupes de cohomologie de la fibre limite.

2. Structure de Hodge mixte limite.

(2.1) Nous revenons aux hypothèses de (1.1) et dans cette situation générale nous allons indiquer comment peuvent se munir d'une structure de Hodge mixte les groupes de cohomologie de la fibre limite.

Soit T l'endomorphisme de monodromie de $H^n(\tilde{X}^*)$ et ρ l'index de quasi-unipotence de T. Si $f': X' \to D$ est la famille obtenue par le changement de base défini par $t = s^\rho$, $\tilde{X}'^*$ la fibre limite correspondante et T' l'endomorphisme de monodromie de $H^n(\tilde{X}'^*)$, il résulte que T' est unipotent. Soit maintenant $Y = f^{-1}(0)$ et $X_\bullet$ une hyperrésolution de $(X', X-Y)$. On rappele que X_α est lisse et que Y_α est un diviseur à croisements normaux dans X_α pour $\alpha \neq 0$. D'après (1.3) il existe un complexe de Hodge mixte cohomologique sur chaque Y_α, A_{X_α}, $\alpha \neq 0$, tel que

$$H^*(\tilde{X}_\alpha^*)_u \cong \mathbb{H}^*(Y_\alpha, A_{X_\alpha}) .$$

Soit

$$A_X = (\mathbb{R}\pi_* A_{X_\bullet^+}, \delta(W, L), F) .$$

On a alors

<u>Théorème</u> [8]. Sous les hypothèses et notations antérieures on a

(i) Les groupes de cohomologie de la fibre limite $H^n(X^*)$ ont une structure de Hodge mixte définie par le complexe $\mathbb{R}\Gamma A_X$.

(ii) La structure de Hodge mixte de $H^n(X^*)$ est indépendante de l'hyperrésolution choisie. Elle est fonctorielle pour les morphismes sur D.

(2.2) Le corollaire qui suit est alors une conséquence des résultats de [12].

<u>Corollaire</u> [8]. Sous les hypothèses et notations antérieures on a

(i) L'endomorphisme N de $H^n(X^*)$ est un morphisme de structures de Hodge mixtes du type $(1, -1)$ qui provient d'un morphisme de complexes de Hodge mixtes cohomologiques.

(ii) Le morphisme $\tilde{X}^* \to X^*$ induit un morphisme de structures de Hodge mixtes $H^n(X^*) \to H^n(\tilde{X}^*)$ qui provient d'un morphisme de complexes de Hodge mixtes cohomologiques.

(iii) La suite de Wang

$$H^n(X^*) \longrightarrow H^n(\tilde{X}^*) \xrightarrow{N} H^n(\tilde{X}^*)(-1)$$

est une suite de structures de Hodge mixtes.

(iv) Le morphisme de spécialisation

$$Sp^*: H^n(Y) \longrightarrow H^n(\tilde{X}^*)$$

est un morphisme de structures de Hodge mixtes.

3. Applications.

(3.1) Le théorème qui suit résoud dans la situation géométrique un problème proposé par Deligne [4].

Théorème [8]. Sous les hypothèses et notations antérieures l'endomorphisme N induit pour tout $b, p, q \geq 0$ des isomorphismes

$$N^b: Gr^W_{q+b}\, Gr^{DecL}_q\, H^{p+q}(X^*) \to Gr^W_{q-b}\, Gr^{DecL}_q\, H^{p+q}(X^*)$$

c'est à dire, la filtration par la monodromie coïncide avec la filtration par le poids W.

Démonstration. On considére la suite spectrale induite par L

$$_L E_1^{pq} = \bigoplus_{|\alpha|=p+1} H^q(X^*_\alpha)_u \Longrightarrow H^{p+q}(X^*) \quad p,\, q \geq 0 .$$

En se spécialisant à une fibre $X_{\alpha,t}$, $t \neq 0$, on voit que cette suite spectrale dégénère en E_2 . D'après [12] on obtient alors l'isomorphisme

$$Gr^W_{q+b}\, E_2^{p,q} \xrightarrow{\sim} Gr^W_{q-b}\, E_2^{p,q}\,(-b)$$

d'où il résulte l'isomorphisme du théorème car

$$E_2^{p,q} \cong E_\infty^{p,q} .$$

Finalement W est determiné par N et L en vertu de [4].

(3.2) Une démonstration analogue à celle qui figure dans [13] (5.12) permet d'obtenir un théorème des cycles invariants et une suite exacte de Clemens-Schmid pour X lisse non nécessairement projective.

RÉFÉRENCES

[1] A. Borel, J. Moore: Homology theory for locally compact spaces, Mich. J., 7 (1960), 137-159.

[2] Clemens, C.H.: Degeneration of Kähler manifolds, Duke Math. J. 44 (1977), 215-290.

[3] P. Deligne: Théorie de Hodge II et III. Publ. Math. I.H.E.S. 40 (1971), 5-58 et 44 (1974), 5-77.

[4] P. Deligne: La conjecture de Weil II. Publ. Math. I.H.E.S., 52 (1980), 137-152.

[5] F. Elzein: Mixed Hodge structures. Trans. A.M.S., 275 (1983), 71-106.

[6] Fujiki: Duality of mixed Hodge structures of algebraic varieties. Publ. RIMS, Kyoto Univ,, 16 (1980), 635-667.

[7] F. Guillén: Une relation entre la filtration par le poids de Deligne et la filtration de Zeeman. A paraitre en Comp. Math.

[8] F.Guillén, V. Navarro Aznar, F. Puerta: Théorie de Hodge via schémas cubiques. Notes polycopiées. Univ. Politècnica de Catalunya, 1982.

[9] F.Guillén, V.Navarro Aznar, P.Pascual Gainza, P.Puerta: Hyperrésolutions cubiques et applications aux théories cohomologiques des variétés algébriques. Prepublication. Univ. Politècnica de Catalunya, 1985.

[10] C. McCrory: Zeeman's filtration of homology. Trans. S.M.S., 250 (1979), 147-166.

[11] C. McCrory: On the topology of Deligne's weight filtration. Proc. Symp. Pure Math., 40 (1983), 217-226.

[12] V. Navarro: Sur la théorie de Hodge-Deligne. Prepublication (1986).

[13] Schmid, W.: Variation of Hodge Structures: The Singularities of the Period Mapping, Invent. Math., 22 (1973), 211-320.

[14] J. Steenbrink: Limits of Hodge Structures. Invent. Math., 31 (1976), 229-257.

[15] J. Steenbrink, J. Stevens: Topological invariance of the weight filtration. Indagat. Math., 46 (1984), 63-76.

[16] J.L. Verdier: Dualité dans la cohomologie des espaces localement compacts. Sem. Bourbaki, 18è. année 65/66, n. 300.

[17] E.C. Zeeman: Dihomology III. A generalitation of the Poincaré duality for manifolds. Proc. London Math. Soc., 3 (13), (1963), 155-183.

Iterated Integrals and Mixed Hodge Structures on Homotopy Groups

Richard M. Hain[1]
Department of Mathematics, GN-50
University of Washington
Seattle, WA 98195
U.S.A.

The aim of this short note is to describe a direct and natural construction of Morgan's mixed Hodge structure on the homotopy groups of a smooth complex algebraic variety [10], using Chen's iterated integrals [2]. This construction is described in [7], but, in presenting the results there in their natural generality, the essential simplicity of the construction has been lost. I hope that this paper exposes that simplicity. I would like to thank Alan Durfee for reading the manuscript and making useful comments.

1. __Strategy.__ First suppose that X is a path connected topological space with each $H_k(X;\mathbb{Q})$ finite dimensional. Choose a base point $x \in X$. We would like a homological method for computing the rational homotopy groups of (X,x).

Associated to (X,x) is the space $P_{x,x}X$ of continuous loops $\gamma : [0,1] \longrightarrow X$ with $\gamma(0) = \gamma(1) = x$. As is well known,

$$\pi_{k+1}(X,x) \cong \pi_k(P_{x,x}X, \eta_x) \qquad k \geq 0 ,$$

where η_x denotes the constant loop at x. Since the path components of $P_{x,x}X$ are the homotopy classes of loops based at x,

$$H_0(P_{x,x}X) \cong \mathbb{Z}\pi_1(X,x) ,$$

the integral group ring of $\pi_1(X,x)$. Alternatively,

$$H^0(P_{x,x}X) \cong \operatorname{Hom}_{\mathbb{Z}}(\mathbb{Z}\pi_1(X,x), \mathbb{Z}) . \qquad (1.1)$$

To state the analogous result for higher homotopy groups we need to recall that

(1.2) a) an __augmented__ __k-algebra__ is a __k-algebra__ A and a homomorphism $\varepsilon : A \longrightarrow k$. (Here k is a commutative ring.) The __augmentation ideal__ I of A is the kernel of ε.

[1] Supported in part by the National Science Foundation through grants MCS-8108814(A04) and DMS-8401175.

b) the <u>space of indecomposables</u> QA of the augmented $\mathscr{k}$-algebra A is the $\mathscr{k}$-module

$$QA = I/I^2 \ .$$

For example, if A is the polynomial ring $\mathscr{k}[x_1,\dots,x_n]$ with the augmentation being evaluation at 0 , then QA is isomorphic to the free $\mathscr{k}$-module generated by $x_1,\dots,x_n$.

Since the cohomology ring of a sphere has no non-trivial cup products, for each topological space Y , the dual of the Hurewicz homomorphism

$$h : H^{\ell}(Y;\mathscr{k}) \longrightarrow \mathrm{Hom}_{\mathbb{Z}}(\pi_{\ell}(Y),\mathscr{k})$$

induces a $\mathscr{k}$-linear map

$$QH^{\ell}(Y;\mathscr{k}) \longrightarrow \mathrm{Hom}(\pi_{\ell}(Y),\mathscr{k}) \ . \qquad (1.3)$$

This map is rarely an isomorphism. (It is not, for example, when $\ell = 3$ and $Y = S^2$.) However, if $\mathscr{k}$ is a field of characteristic zero and Y is a connected H-space, then the Borel-Serre theorem [9] guarantees that (1.3) is an isomorphism. When X is simply connected, $P_{x,x}X$ is such a space. Thus we have the following version of the Borel-Serre theorem:

(1.4) <u>Theorem.</u> If X is simply connected, then the natural map

$$QH^{\ell}(P_{x,x}X\,;\,\mathbb{Q}) \longrightarrow \mathrm{Hom}_{\mathbb{Z}}(\pi_{\ell+1}(X,x)\,,\,\mathbb{Q})$$

is an isomorphism. □

Actually, each side has the structure of a graded Lie coalgebra (i.e., the dual of each side is a graded Lie algebra) and the isomorphism preserves this structure.

Thus one approach to computing rational homotopy groups homologically would be to find a complex that computes the indecomposables of loop space cohomology. This can be done using the bar construction and its geometric manifestation, iterated integrals.

At this point it is worth noting that multiplication of paths

$$P_{x,x}X \times P_{x,x}X \longrightarrow P_{x,x}X$$

induces an algebra homomorphism

$$\Delta : H^{\bullet}(P_{x,x}X\,;\,\mathbb{Q}) \longrightarrow H^{\bullet}(P_{x,x}X\,;\,\mathbb{Q}) \otimes H^{\bullet}(P_{x,x}X\,;\,\mathbb{Q}) \ .$$

Together with the cup product, this gives $H^{\bullet}(P_{x,x}X\,;\,\mathbb{Q})$ the structure of a graded Hopf algebra (see [9]).

2. Iterated Integrals. Suppose that M is a smooth manifold and that $w_1,\dots,w_r$ are differential forms on M, each of degree ≥ 1. The iterated integral $\int w_1\dots w_r$ is a differential form on PM, the space of piecewise smooth paths in M, of degree

$$\sum_{j=1}^{r} (\deg w_j - 1) \tag{2.1}$$

To make this meaningful, we need the notion of a smooth map $\alpha : N \longrightarrow PM$ from a manifold N into PM: The map α corresponds to a map

$$\phi_\alpha : [0,1] \times N \longrightarrow M$$
$$(t,\xi) \longmapsto \alpha(\xi)(t) .$$

We say that α is smooth if it is continuous and there is a partition $0 = t_0 < t_1 < \dots < t_m = 1$ of $[0,1]$ such that the restriction of ϕ_α to each $[t_{j-1}, t_j] \times N$ is smooth.

An n-form w on PM is given by specifying its pullback $\alpha^* w \in E^n N$ (= n-forms on N) along each smooth map $\alpha : N \longrightarrow PM$. (Here N ranges over all smooth manifolds.) These pullbacks are required to satisfy the obvious compatibility condition. Namely, if $\theta : W \longrightarrow N$ is smooth, then

$$\theta^*(\alpha^* w) = (\alpha \circ \theta)^* w .$$

(For full details including the definitions of exterior differentiation and wedge product, see [2].)

To define $\alpha^* \int w_1 \dots w_r$, write

$$\phi_\alpha^* w_j = w_j'(t,\xi) + dt \wedge w_j''(t,\xi) ,$$

where w', w'' contain no dt's. (That is, w' is the interior product of $\phi_\alpha^* w$ with the vector field $\partial/\partial t$.) Now we define

$$\alpha^* \int w_1 \dots w_r = \int_{0 < t_1 \leq \dots \leq t_r < 1} \dots \int w_1'(t_1,\xi) \wedge \dots \wedge w_r'(t_r,\xi) dt_1 dt_2 \dots dt_r .$$

This is a differential form on N of degree given by (2.1).

The linear subspace of the de Rham complex of PM spanned by the iterated integrals $\int w_1 \dots w_r$, where each w_j lies in a fixed d.g. sub algebra $A^\bullet$ of the de Rham complex of M, forms a d.g. algebra that we shall denote by $\int A^\bullet$. This is proved, for example, in [2]. If $x \in M$, then we can restrict each iterated integral to the loop space $P_{x,x} M$. Denote this complex of iterated integrals by $\int A_x^\bullet$. It has the structure of a d.g. Hopf algebra with coproduct dual to the multiplication of paths.

3. The Bar Construction. Let $A^\bullet$ be a sub complex of the de Rham complex of M, a smooth manifold. Fix a base point $x \in M$. The bar construction $B(A)_x$ essentially gives an algebraic description of $\int A^\bullet_x$.

Since the de Rham complex of a point is $\mathbb{R}$, the inclusion of x into M induces an augmentation

$$\varepsilon : A^\bullet \longrightarrow \mathbb{R} .$$

Denote the corresponding augmentation ideal $\ker \varepsilon$ by $IA^\bullet$. The bar construction $B(A^\bullet)_x$ on $A^\bullet$ is the total complex associated to the double complex

$$(B^{-s,t}(A^\bullet), d', d'') ,$$

where

$$B^{-s,t} = (\otimes^s IA)^t .$$

The right hand side denotes the degree t piece of $\otimes^s IA$. It is traditional to denote the element $w_1 \otimes \dots \otimes w_s$ of $B^{-s,\bullet}$ by $[w_1|\dots|w_s]$. The differentials

$$d' : B^{-s,t} \longrightarrow B^{-s+1,t} , \quad d'' : B^{-s,t} \longrightarrow B^{-s,t+1}$$

are defined by

$$d'[w_1|\dots|w_r] = \sum_{i=1}^{r-1} (-1)^{i+1}[Jw_1|\dots|Jw_{i-1}|Jw_i \wedge w_{i+1}|w_{i+2}|\dots|w_r] \tag{3.1}$$

and

$$d''[w_1|\dots|w_r] = \sum_{i=1}^{r} (-1)^{i}[Jw_1|\dots|Jw_{i-1}|dw_i|\dots|w_r] . \tag{3.2}$$

Here Jw denotes $(-1)^{\deg w} w$.

The bar construction $B(A^\bullet)_x$ has the structure of a d.g. Hopf algebra. The coproduct $\Delta : B(A^\bullet)_x \longrightarrow B(A^\bullet)_x \otimes B(A^\bullet)_x$ is induced by

$$[w_1|\dots|w_r] \longmapsto \sum_{i=0}^{r} [w_1|\dots|w_i] \otimes [w_{i+1}|\dots|w_r] \tag{3.3}$$

and the product $B(A^\bullet)_x \otimes B(A^\bullet)_x \longrightarrow B(A^\bullet)_x$ by

$$[w_1|\dots|w_r] \otimes [w_{r+1}|\dots|w_{r+s}] \longmapsto \sum_{\sigma} \varepsilon(\sigma)[w_{\sigma(1)}|\dots|w_{\sigma(r+s)}] , \tag{3.4}$$

where σ ranges over all shuffles of type (r,s) and $\varepsilon : S_{r+s} \longrightarrow \{-1,1\}$ is the weighted sign representation of the symmetric group obtained by assigning weight $-1+\deg w_j$ to w_j. One should note that, although $B(A^\bullet)$ makes sense for any augmented d.g. algebra $A^\bullet$, the product only commutes with the differential when $A^\bullet$ is commutative (in the graded sense).

By defining

$$\underline{\underline{B}}^{-s}B(A^\bullet)_x = \bigoplus_{u \le s} B^{-u,\bullet} , \tag{3.5}$$

we obtain a decreasing filtration of $B(A^\bullet)_x$ and hence a spectral sequence. This spectral sequence is an instance of the Eilenberg-Moore spectral sequence. Its E_1 term is $B(HA^\bullet)$. From this we conclude that the bar construction takes quasi-isomorphisms of augmented algebras to quasi-isomorphisms.

(3.6) Theorem (Chen [2]). a) The linear map

$$B(A^\bullet)_x \longrightarrow \int A^\bullet_x \tag{3.7}$$

$$[w_1|\dots|w_r] \longmapsto \int w_1 \dots w_r$$

is a d.g. Hopf algebra homomorphism.

b) If the inclusion $A^\bullet \longrightarrow E^\bullet M$ of $A^\bullet$ into the de Rham complex of M is a quasi-isomorphism, then (3.7) is also a quasi-isomorphism. □

Standard arguments from algebraic topology, dating back to Adams [1] in the simply connected case and Chen [2] and Stallings [11] for the fundamental group, can then be used to prove the following results.

(3.8) Theorem (Adams-Chen). If M is simply connected and the inclusion $A^\bullet \longrightarrow E^\bullet M$ is a quasi-isomorphism, then the integration map

$$H^\bullet(\int A^\bullet_x) \longrightarrow H^\bullet(P_{x,x}M;\mathbb{R})$$

is a Hopf algebra isomorphism. □

The group ring $k\pi_1(M,x)$ has a natural augmentation $k\pi_1(M,x) \longrightarrow k$ induced by the constant map $M \longrightarrow \{x\}$. Denote the corresponding augmentation ideal by J.

(3.9) Theorem (Chen, Stallings). If the inclusion $A^\bullet \longrightarrow E^\bullet M$ induces an isomorphism on H^1 and an injection on H^2, then

a) integration induces a pairing of Hopf algebras

$$\langle , \rangle : H^0(\int A^\bullet_x) \otimes \mathbb{R}\pi_1(M,x) \longrightarrow \mathbb{R}$$

such that

$$\langle \underline{\underline{B}}_s H^0(\int A^\bullet_x), J^{s+1} \rangle = 0 .$$

b) The map

$$\underline{\underline{B}}_s H^0(\int A^\bullet_x) \longrightarrow \mathrm{Hom}_{\mathbb{Z}}(\mathbb{Z}\pi_1(M,x)/J^{s+1},\mathbb{R})$$

is an isomorphism. □

One should note that, as there are no iterated integrals of degree < 0 (they are differential forms),

$$\underline{\underline{B}}_s H^0(\int A_x^\bullet) = H^0(\text{iterated integrals of length } \leq s).$$

An elementary proof of (3.9) that gives insights into the geometry of iterated integrals can be found in [8].

Finally, one should note that (3.7) is very close to being an isomorphism. If $H^1(A^\bullet) \longrightarrow H^1(M)$ is injective, then the kernel of (3.7) is spanned by elements of the following types:

$$[w_1|\dots|w_r] \text{ , where at least one } w_j \in A^0 , \tag{3.10)(i}$$

$$[w_1|\dots|w_i|df|w_{i+1}|\dots|w_r] - [w_1|\dots|w_i|fw_{i+1}|\dots|w_r] + [w_1|\dots|fw_i|w_{i+1}|\dots|w_r] , \quad \text{where} \quad f \in A^0 , \tag{ii}$$

$$[w_1|\dots|w_r|df] - f(x)[w_1|\dots|w_r] + [w_1|\dots|fw_r] , \quad \text{where} \quad f \in A^0 , \tag{iii}$$

$$[df|w_1|\dots|w_r] - [fw_1|\dots|w_r] + f(x)[w_1|\dots|w_r] , \quad \text{where} \quad f \in A^0 . \tag{iv}$$

That these relations hold in $\int A_x^\bullet$ can be verified using integration by parts. Chen has shown that these are the only relations in $\int A_x^\bullet$. In general, Chen has defined the reduced bar construction $\overline{B}(A^\bullet)_x$ on an augmented d.g. algebra $A^\bullet$ to be $B(A^\bullet)$ modulo the relations (3.10) (see [3]). (In this case $f(x)$ is to be interpreted as $\varepsilon(f)$.) This gives a purely algebraic description of iterated integrals that is useful in Hodge theory.

4. Mixed Hodge Complexes. We assume that the reader is familiar with Deligne's notion of a mixed Hodge complex (MHC) [4:(8.1.5)]. We extend this notion by defining a multiplicative MHC to be a MHC

$$\underline{\underline{A}} = ((A_{\mathbb{Q}}^\bullet, W_\bullet) , (A_{\mathbb{C}}^\bullet, W_\bullet, F^\bullet))$$

where

(i) $A_{\mathbb{Q}}^\bullet$ and $A_{\mathbb{C}}^\bullet$ are d.g. algebras,

(ii) $A_{\mathbb{Q}}^\bullet \otimes \mathbb{C}$ and $A_{\mathbb{C}}^\bullet$ are $W_\bullet$ filtered quasi-isomorphic as d.g. algebras,

(iii) all filtrations are preserved by the products.

This can be generalized to define augmented multiplicative MHC's in the obvious way. To an augmented multiplicative MHC we can associate its bar construction

$$B(\underline{\underline{A}}) = (B(A_{\mathbb{Q}}^\bullet), W_\bullet * \underline{\underline{B}}_\bullet) , (B(A_{\mathbb{C}}^\bullet), W_\bullet * \underline{\underline{B}}_\bullet , F^\bullet)) ,$$

where $W_\bullet$ and $F^\bullet$ are the natural extensions of the Hodge and weight filtrations to $B(\underline{\underline{A}})$ and $W_\bullet * \underline{\underline{B}}_\bullet$ is the convolution of the bar and weight filtrations

$$(W_\bullet * \underline{\underline{B}}_\bullet)_\ell = \bigoplus_{s \geq 0} W_{\ell - s} \otimes^s I\underline{\underline{A}}^\bullet \quad .$$

The basic result we need for putting a mixed Hodge structure (MHS) on homotopy groups is the following.

(4.1) <u>Lemma</u>. If $\underline{\underline{A}}$ is an augmented multiplicative MHC, then $B(\underline{\underline{A}})$ is a MHC and the coproduct $B(\underline{\underline{A}}) \longrightarrow B(\underline{\underline{A}}) \otimes B(\underline{\underline{A}})$ preserves the filtrations. If $\underline{\underline{A}}$ is commutative, then the shuffle product $B(\underline{\underline{A}}) \otimes B(\underline{\underline{A}}) \longrightarrow B(\underline{\underline{A}})$ preserves the filtrations.

The proof is an easy exercise using the definitions. To get started, note that

$$_{W * \underline{\underline{B}}}E_0^{\ell, m}(B(\underline{\underline{A}})) = \bigoplus_s {}_W E_0^{\ell + s, m}(\otimes^s I\underline{\underline{A}}) \quad .$$

Combining (4.1) with the result of [6], we obtain a MHC for homotopy.

(4.2) <u>Corollary</u>. If $\underline{\underline{A}}$ is a commutative, augmented, multiplicative MHC, then $QB(\underline{\underline{A}})$ is a MHC. □

For later reference, note that another consequence of the main result of [6] is that, for an augmented commutative d.g. algebra $A^\bullet$, there is a natural Lie coalgebra isomorphism

$$QHB(A^\bullet) \cong HQB(A^\bullet) \quad . \qquad (4.3)$$

5. <u>Mixed Hodge Structures on Homotopy Groups</u>. Suppose that V is a smooth algebraic variety over $\mathbb{C}$ (or a Zariski open subset of a compact Kähler manifold). Choose a smooth projective completion X of V such that $X - V$ is a divisor D in X with normal crossings. The C^∞ log complex $E^\bullet(X \log D)$ is the complex part $A^\bullet_{\mathbb{C}}$ of a multiplicative MHC

$$\underline{\underline{A}} = ((A^\bullet_{\mathbb{R}}, W_\bullet), (A^\bullet_{\mathbb{C}}, W_\bullet, F^\bullet))$$

that computes the real cohomology of V. The real part $A^\bullet_{\mathbb{R}}$ of this MHC is a real analogue of the log complex (see, for example, [5]) — if D is locally given by $z_1 z_2 \ldots z_k = 0$, then $A^\bullet_{\mathbb{R}}$ is locally generated by real valued forms on X and the forms

$$d\theta_j = \frac{1}{4\pi i}\left(\frac{dz_j}{z_j} - \frac{d\bar{z}_j}{\bar{z}_j}\right) \qquad j = 1, \ldots, k \quad .$$

The weight filtration on $A^\bullet_{\mathbb{R}}$ is defined by giving each $d\theta_j$ weight 1 and extending multiplicatively.

Since $A^{\bullet}_{\mathbb{R}}$ and $A^{\bullet}_{\mathbb{C}}$ are sub d.g. algebras of $E^{\bullet}V$, we may combine (3.8), (3.9), (4.1) and (4.2) to obtain the desired MHS on homotopy.

(5.1) <u>Theorem</u>. If V is a smooth algebraic variety over $\mathbb{C}$ and $x \in V$, then, for each $s \geq 0$, $\mathbb{Z}\pi_1(V,x)/J^{s+1}$ has a natural MHS. If V is simply connected, then $\pi_{\bullet}(V,x)$ has a MHS. These MHS's are functorial with respect to base point preserving morphisms of varieties and, moreover, all algebraic operations preserve the MHS. □

As we have described (5.1) the weight filtration is only defined over $\mathbb{R}$. With more care and greater technical complication, one can show that $W_{\bullet}$ is defined over $\mathbb{Q}$. Also, there is nothing special about smooth varieties — (5.1) holds for all complex algebraic varieties. The details may be found in [7].

Often in Hodge theory it is useful to know that the maps and long exact sequences of algebraic topology are compatible with Hodge theory. In this context one can prove (see [7]):

(5.2) <u>Theorem</u>. a) If W is a subvariety of the algebraic variety V and if both V and W are simply connected, then the relative homotopy groups $\pi_{\bullet}(V,W,x)$ have a natural MHS and all maps in the diagram

$$\begin{array}{ccccccccccc}
\cdots \longrightarrow & \pi_i(W,x) & \longrightarrow & \pi_i(V,x) & \longrightarrow & \pi_i(V,W,x) & \longrightarrow & \pi_{i-1}(W,x) & \longrightarrow \cdots \\
& \downarrow & & \downarrow & & \downarrow & & \downarrow & \\
\cdots \longrightarrow & H_i(W) & \longrightarrow & H_i(V) & \longrightarrow & H_i(V,W) & \longrightarrow & H_{i-1}(W) & \longrightarrow \cdots
\end{array}$$

are morphisms of MHS. In particular the Hurewicz homomorphism is a morphism.

b) If $f: X \longrightarrow Y$ is a fiber bundle of simply connected complex algebraic varieties, then the long exact sequence of homotopy groups

$$\cdots \longrightarrow \pi_i(X_y,x) \longrightarrow \pi_i(X,x) \longrightarrow \pi_i(Y,y) \longrightarrow \pi_{i-1}(X_y) \longrightarrow \cdots$$

is a long exact sequence of MHS's. Here X_y denotes the fiber over y and $f(x) = y$. □

Our final remarks concern the dependence of the MHS on $\pi_{\bullet}(V,x)$ upon the base point x. If V is not simply connected, then the MHS on $\mathbb{Z}\pi_1(V,x)/J^{s+1}$ may depend non trivially on x (see [8]). When V is simply connected, $\pi_{\bullet}(V,x)$ does not vary when x is varied. Although it is not immediately obvious, the same holds for the MHS on $\pi_{\bullet}(V,x)$. That is, the canonical isomorphism $\pi_{\bullet}(V,x) \longrightarrow \pi_{\bullet}(V,y)$ is an isomorphism of MHS. These assertions are proved in [7].

References

[1] Adams, J. F., On the cobar construction, Colloque de Topologie Algébrique (Louvain, 1956), George Thone, Paris, 1957, 81-87.

[2] Chen, K.-T., Iterated path integrals, Bull. Amer. Math. Soc., 83 (1977), 831-879.

[3] Chen, K.-T., Reduced bar constructions on de Rham complexes. In: Heller, A., Tierney, M., eds., Algebra, Topology, and Category Theory, Academic Press, New York, 1976, 19-32.

[4] Deligne, P., Théorie de Hodge, III. Publ. Math. IHES 44, (1974), 5-77.

[5] Durfee, A., Hain, R., Mixed Hodge structures on the homotopy of links. To appear.

[6] Hain, R., On the indecomposables of the bar construction, Proc. Amer. Math. Soc., to appear.

[7] Hain, R., The de Rham homotopy theory of complex algebraic varieties, I. To appear.

[8] Hain, R., The geometry of the mixed Hodge structure on the fundamental group. To appear in Proc. of the AMS Summer Institute, Algebraic Geometry, Bowdoin College, 1985. Proc. Symp. Pure Math.

[9] Milnor, J., Moore, J., On the structure of Hopf algebras, Ann. Math, 81 (1965), 211-264.

[10] Morgan, J., The algebraic topology of smooth algebraic varieties, Publ. IHES, 48 (1978), 137-204.

[11] Stallings, J., Quotients of the powers of the augmentation ideal in a group ring. In Knots, Groups and 3-Manifolds, Papers Dedicated to the Memory of R. H. Fox, L. Neuwirth ed., Princeton University Press, 1975.

Higher Albanese Manifolds

Richard M. Hain[1]
Department of Mathematics, GN-50
University of Washington
Seattle, WA 98195
U.S.A.

The classical albanese variety of a smooth projective variety X is the complex torus

$$\mathrm{Alb}\, X := \Omega^1(X)^* / H_1(X\,;\mathbb{Z}) .$$

Choosing a base point $x \in X$, one obtains a holomorphic mapping

$$\alpha_x : X \longrightarrow \mathrm{Alb}\, X ,$$
$$y \longmapsto \int_x^y$$

which is called the albanese mapping. The albanese variety is an Eilenberg-MacLane space with abelian fundamental group:

$$\pi_k(\mathrm{Alb}\, X) = \begin{cases} H_1(X)/\text{torsion} & k=1 \\ 0 & k>1 \end{cases} .$$

In this note we define higher albanese manifolds of a smooth complex algebraic variety and give an explicit formula for the generalized albanese mappings. The generalized albanese manifolds of the variety X form an inverse system

$$\cdots \longrightarrow \mathrm{Alb}^3 X \longrightarrow \mathrm{Alb}^2 X \longrightarrow \mathrm{Alb}^1 X$$

of complex manifolds and holomorphic maps. There is a sequence of compatible holomorphic maps

$$\alpha^s : X \longrightarrow \mathrm{Alb}^s X$$

that are natural with respect to morphisms of smooth varieties. Let

$$\pi_1(X,x) = \Gamma^1 \geq \Gamma^2 \geq \cdots$$

denote the lower central series of $\pi_1(X,x)$. (i.e., $\Gamma^{s+1} = [\Gamma^1, \Gamma^s]$.) The generalized albanese manifolds are Eilenberg-MacLane spaces:

$$\pi_k(\mathrm{Alb}^s X) = \begin{cases} \pi_1(X,x)/\Gamma^{s+1}\,)/\text{torsion} & k=1 \\ 0 & k>1 \end{cases} .$$

[1] Supported in part by the National Science Foundation through grants MCS-8108814(A04) and DMS-8401175.

Unfortunately these generalized Albaneses are rarely algebraic varieties (or even Zariski open subsets of compact Kähler manifolds)(see [7]), but they do arise naturally in the classification of unipotent variations of mixed Hodge structure over X . Their construction was previously known to Deligne [3]. Prior to this, Parsin [9] had defined a sequence of generalized albanese mappings for a compact Riemann surface. These were later studied by Hwang-Ma [8]. However, Parsin's construction does not yield higher albanese manifolds for Riemann surfaces because the family of multi-valued functions he used to construct the albanese mapping was too small — the monodromy group was, in general, not discrete. Chen [1] has constructed a C^∞ analogue of the higher albanese mappings for compact Riemannian manifolds.

The material in this paper is an elaboration of a small part of joint work [7] with S. Zucker.

1. Higher Albanese Manifolds. Let V be a smooth complex algebraic variety and $x \in V$. Let s be a positive integer. The complex form of the Malcev completion of $\pi_1(V,x)/\Gamma^{s+1}$ is a simply connected, complex, nilpotent Lie group $G_s(\mathbb{C})$ together with a group homomorphism

$$\theta_x^s : \pi_1(V,x)/\Gamma^{s+1} \longrightarrow G_s(\mathbb{C}) \quad . \tag{1.1}$$

Let $G_s(\mathbb{Z})$ denote the image of θ_x^s. The Malcev completion (1.1) is characterized by the following properties:

(1.2) a) The kernel of θ_x^s is a finite group.

b) If

$$G_s(\mathbb{C}) = \Gamma^1 G_s \supseteq \Gamma^2 G_s \supseteq \cdots$$

denotes the lower central series of $G_s(\mathbb{C})$, then, for $1 \le t \le s$,

$$[\mathrm{gr}_\Gamma^t \pi_1(V,x)] \otimes \mathbb{C} \cong \mathrm{gr}_\Gamma^t G_s(\mathbb{C}) \quad .$$

For example, when $s = 1$ we have

$$G_1(\mathbb{C}) = H_1(V;\mathbb{C})$$

and θ_x^1 is the composite

$$\pi_1(V,x)/\Gamma = H_1(V;\mathbb{Z}) \longrightarrow H_1(V,\mathbb{C}) = G_1(\mathbb{C}) \quad .$$

Observe that, when V is projective, the Hodge filtration on $H_1(V)$ satisfies

$$H_1(V;\mathbb{C}) = F^{-1} \supseteq F^0 \supseteq F^1 = 0$$

and

$$H_1(V;\mathbb{C})/F^0 \cong \Omega^1(V)^* \quad .$$

Consequently,

$$\mathrm{Alb}\,V = G_1(\mathbb{Z})\backslash G_1(\mathbb{C})/F^0G_1(\mathbb{C}) \quad .$$

To generalize this construction to $s > 1$ we need to introduce the mixed Hodge structure (MHS) on $\pi_1(V,x)$.

The Malcev group $G_s(\mathbb{C})$ and its Lie algebra $\mathfrak{g}_s(\mathbb{C})$ can be realized inside the truncated group ring $\mathbb{C}\pi_1(V,x)/J^{s+1}$:

$$\begin{array}{ccc} \mathfrak{g}_s(\mathbb{C}) & \underset{\log}{\overset{\exp}{\rightleftarrows}} & G_s(\mathbb{C}) \\ & \searrow \quad \swarrow & \\ & \mathbb{C}\pi_1(V,x)/J^{s+1} & \end{array} \qquad (1.3)$$

Details of this construction, due to Quillen [10:Appendix A], can be found, for example, in [7:§2].

The Lie algebra $\mathfrak{g}_s(\mathbb{C})$ is a subspace of $\mathbb{C}\pi_1(V,x)/J^{s+1}$ defined over $\mathbb{Q}$. Its bracket is the restriction of the commutator $[A,B] = AB - BA$ to $\mathfrak{g}_s(\mathbb{C})$. It follows that $\mathfrak{g}_s(\mathbb{C})$ has a MHS and that the bracket is a morphism of MHS. Since the bracket preserves the Hodge filtration, $F^0\mathfrak{g}_s$ is a sub Lie algebra of $\mathfrak{g}_s(\mathbb{C})$. Denote the corresponding closed subgroup of $G_s(\mathbb{C})$ by F^0G_s . Because $G_s(\mathbb{C})$ and F^0G_s are contractible, so is $G_s(\mathbb{C})/F^0G_s$. Since $Gr_\ell^W\mathfrak{g}_s = 0$ when $\ell \geq 0$, $F^0\mathfrak{g} \cap \overline{F}^0\mathfrak{g} = 0$ from which it follows that $G_s(\mathbb{Z})$ acts freely and discontinuously on $G_s(\mathbb{C})/F^0G_s$.

(1.4) <u>Definition</u>. The sth Albanese manifold, Alb_x^sV , of (V,x) is defined by

$$\mathrm{Alb}_x^sV = G_s(\mathbb{Z})\backslash G_s(\mathbb{C})/F^0G_s \quad .$$

(1.5) <u>Remarks</u>. Suppose that $V = X - D$ where X is a smooth complete variety and D is a divisor in X with normal crossings.

a) If $h^{1,0}(X) = 0$, then $F^0\mathfrak{g}_s = 0$ and

$$\mathrm{Alb}_x^sV = G_s(\mathbb{Z})\backslash G_s(\mathbb{C}) \quad .$$

b) Since $G_s(\mathbb{C})/F^0G_s$ is contractible, Alb_x^sV is an Eilenberg-MacLane space $K(G_s(\mathbb{Z}),1)$.

c) When $s = 1$

$$\mathrm{Alb}_x^1V \cong \Omega^1(X \log D)^*/H_1(V;\mathbb{Z}) \quad .$$

d) It is not immediately clear that Alb_x^sV is independent of the base point x , which it is.

2. <u>Higher Albanese Mappings: a Special Case.</u> Suppose that $V = X - D$ where X is a smooth complete variety and D is a divisor in X with normal crossings. Define the <u>irregularity</u> $q(V)$ of V by

$$q(V) = \frac{1}{2}\dim W_1 H^1(V;\mathbb{C}) = h^{1,0}(X) .$$

In this section we construct the Albanese mapping α_x^s for varieties with irregularity 0. In the next section we will sketch the construction of α_x^s in the general case.

Suppose that $q(V) = 0$. As noted in (1.7)a), this implies that $F^0\mathcal{G}_s = 0$, so that

$$\mathrm{Alb}_x^s V = G_s(\mathbb{Z}) \backslash G_s(\mathbb{C}) . \tag{2.1}$$

To give a formula for α_x^s we first need an explicit description of $G_s(\mathbb{C})$.

Since $q(V) = 0$, the inclusion

$$\Omega^\bullet(X \log D) \longrightarrow E^\bullet V$$

of the global holomorphic differentials on V with logarithmic singularities along D into the C^∞ de Rham complex of V induces an isomorphism on H^1 and an injection on H^2. Thus, by [5:(3.9)], integration induces an isomorphism

$$\underline{\underline{B}}_s H^0(B(\Omega^\bullet(X \log D)))^* \cong \mathbb{C}\pi_1(X,x)/J^{s+1} . \tag{2.2}$$

(For notation, see [5].) To get an explicit presentation of $G_s(\mathbb{C})$, choose bases $w_1,\ldots,w_m$ of $\Omega^1(X \log D)$, $z_1,\ldots,z_n$ of $\Omega^2(X \log D)$. Let $X_1,\ldots,X_m$ be the basis of $H_1(V;\mathbb{C}) \cong \Omega^1(X \log D)^*$ dual to the w_j.

Denote the free associative algebra generated by the intermediates X_j by $\mathbb{C}\langle X_1,\ldots,X_m\rangle$. Let I be the ideal generated by the X_j. Define an algebra homomorphism

$$T : \mathbb{C}\langle X_1,\ldots,X_m\rangle \longrightarrow H^0(B(\Omega^\bullet(X \log D)))^*$$

by defining

$$\langle X_i, [w_{j_1}|\ldots|w_{j_r}]\rangle = \begin{cases} 0 & r > 1 \\ \langle X_i, w_{j_1}\rangle & r = 1 . \end{cases}$$

(2.3) <u>Proposition.</u> T induces an isomorphism

$$\mathbb{C}\langle X_1,\ldots,X_m\rangle/\Big(\sum_{i,j} a_{ij}^k [X_i,X_j], k = 1,\ldots,n\Big) + I^{s+1} \cong \underline{\underline{B}}_s H^0(B(\Omega^\bullet(X \log D)))^* ,$$

where the a_{ij}^k are the unique complex constants such that

$$w_i \wedge w_j = \sum a_{ij}^k z_k .$$

The proof is an easy exercise using the definition of the bar construction (see, for example, [5:§3]) and the fact that each element of $\Omega^\bullet(X \log D)$ is closed and none is exact (see [2:(3.2.14)]).

Combining this with (2.2), we obtain:

(2.4) <u>Corollary</u>. If $q(V) = 0$, then there is an algebra isomorphism

$$\theta_x^s : \mathbb{C}\pi_1(V,x)/J^{s+1} \xrightarrow{\approx} \mathbb{C}\langle X_1,\ldots,X_m\rangle / (\Sigma\, a_{ij}^k [X_i,X_j]) + I^{s+1}$$

which is natural with respect to morphisms of smooth varieties. □

An explicit formula for θ_x^s can be obtained as follows. The isomorphism (2.2) takes the homotopy class of the path γ to the functional on $H^0(B(\Omega^\bullet(X \log D)))$ induced by the functional on $B(\Omega^\bullet(X \log D))$ that takes $[w_{i_1}|\ldots|w_{i_r}]$ to $\int_\gamma w_{i_1}\ldots w_{i_r}$. Thus, θ_x^s takes γ to the element

$$1 + \sum_{1\le r\le s} \int_\gamma w_{i_1}\ldots w_{i_r} X_{i_1}\ldots X_{i_r} \tag{2.5}$$

of $A_s = \mathbb{C}\langle X_1,\ldots,X_m\rangle/\text{relns}$. At this stage it is convenient to let

$$T_s = 1 + \Sigma \int w_i X_i + \Sigma \int w_i w_j X_i X_j + \cdots \tag{2.6}$$

This is an A_s-valued iterated integral on V. One can rewrite the formula (2.5) for θ_x^s as

$$\theta_x^s(\gamma) = \langle T,\gamma\rangle \quad . \tag{2.7}$$

The imbeddings (1.3) of $\mathfrak{g}_s(\mathbb{C})$ and $G_s(\mathbb{C})$ into A_s can be realized explicitly as follows: View A_s as a Lie algebra with the commutator as bracket. Then $\mathfrak{g}_s(\mathbb{C})$ is the sub Lie algebra generated by $X_1,\ldots,X_m$ and

$$G_s(\mathbb{C}) = \{\exp U : U \in \mathfrak{g}_s(\mathbb{C})\} \quad .$$

This can be proven by combining [5:(3.9)] with Quillen's construction of $G_s(\mathbb{C})$ [10:Appendix A].

The final fact that we shall need for constructing α_x^s is the next result.

(2.8) <u>Proposition</u>. If γ is a path in V, then $\langle T,\gamma\rangle \in G_s(\mathbb{C})$.

<u>Proof</u>. Set $\omega = w_1X_1 + \cdots + w_mX_m$. This is an element of $\Omega^1(X \log D) \otimes \mathfrak{g}_s(\mathbb{C})$. Observe that

$$T = 1 + \int\omega + \int\omega\omega + \int\omega\omega\omega + \cdots \quad .$$

Let $\gamma_t : [0,1] \longrightarrow V$ be the path defined by $\gamma_t(s) = \gamma(st)$. Since γ_0 is a constant path, $T(\gamma_0) = 1$. Since

$$\frac{d}{dt}\langle T,\gamma_t\rangle = \langle T,\gamma_t\rangle \; \langle\omega,\dot\gamma(t)\rangle \;, \tag{2.9}$$

the function $t \longmapsto \langle T, \gamma_t \rangle$ satisfies the equation

$$X'(t) = X(t)A(t) \ , \quad X(0) = 1$$

in A_s, where $A(t) \in \mathfrak{g}_s(\mathbb{C})$. The result now follows from basic Lie theory. $\square$

The albanese mapping

$$\alpha_x^s : V \longrightarrow \mathrm{Alb}_x^s V \ ,$$

is now defined by

$$\alpha_x^s(z) = \langle T, \gamma \rangle \ , \tag{2.10}$$

where γ is any path in V from x to z. From (2.1) it follows that α_x^s is well defined and from (2.9) that it is holomorphic.

Next we establish independence from the basepoint.

(2.11) <u>Proposition.</u> If $x, y \in V$, then there is a canonical holomorphic map

$$\tau_x^y : \mathrm{Alb}_x^s V \longrightarrow \mathrm{Alb}_y^s V$$

such that the diagram

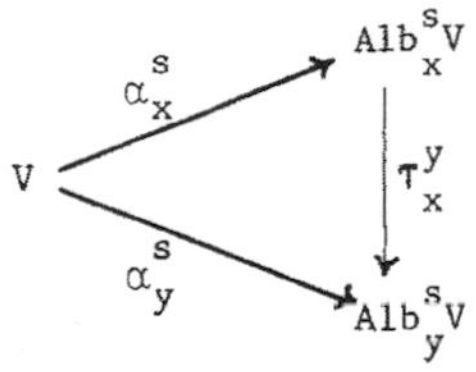

commutes.

<u>Proof.</u> Choose any path γ from y to x in V. Then τ_x^y is induced by left multiplication by $\langle T, \gamma \rangle$ in $G_s(\mathbb{C})$. It is an exercise to show that this induces a well defined map of albanese manifolds. $\square$

Since we have such canonical isomorphisms, the albanese manifold and map are independent of the base point and will be henceforth written

$$\alpha^s : V \longrightarrow \mathrm{Alb}^s V \ . \tag{2.12}$$

We leave the proof of the next result, which establishes the naturality of the higher albanese mappings, to the reader.

(2.13) <u>Proposition.</u> Suppose that $f : V \longrightarrow W$ is a morphism between smooth varieties with $q = 0$, then there is a map

$$\mathrm{Alb}^s(f) : \mathrm{Alb}^s V \longrightarrow \mathrm{Alb}^s W$$

such that

$$\begin{array}{ccc} V & \xrightarrow{\alpha^s} & \mathrm{Alb}^s V \\ {\scriptstyle f}\downarrow & & \downarrow{\scriptstyle \mathrm{Alb}(f)} \\ W & \xrightarrow[\alpha^s]{} & \mathrm{Alb}^s W \end{array}$$

commutes. □

3. Higher Albanese Mappings: the General Case. Suppose that V is a smooth complex algebraic variety. The Lie algebra $\mathfrak{g}_s$ associated with $\pi_1(V,x)$ carries a functorial MHS such that the inclusion (1.3) of $\mathfrak{g}_s$ into $\mathbb{C}\pi_1(V,x)/J^{s+1}$ is a morphism of MHS. Let

$$\mathfrak{g}_s(\mathbb{C}) = \oplus\, \mathfrak{g}_s^{p,q}$$

be a complex splitting of the Hodge and weight filtrations. That is,

$$W_\ell \mathfrak{g}_s(\mathbb{C}) = \bigoplus_{p+q\leq \ell} \mathfrak{g}_s^{p,q}\ , \quad F^p \mathfrak{g}_s(\mathbb{C}) = \bigoplus_{u\geq p} \mathfrak{g}_s^{u,\bullet}\ .$$

Choose a bigraded basis $\{U_j\}$ of $\mathfrak{g}_s(\mathbb{C})$. One can construct, using Chen's method of power series connections (see [6]), a $\mathfrak{g}_s(\mathbb{C})$-valued 1-form

$$\omega \in E^1(X \log D) \otimes \mathfrak{g}_s(\mathbb{C})$$

satisfying

(3.1)a) $\quad d\omega + \frac{1}{2}[\omega,\omega] = 0$

b) $\quad$ if $\omega = \sum w_j U_j$ and

$U_j \in \mathfrak{g}^{-p,-\ell+p}$, then $w_j \in F^p \cap W_{\ell-1}E^1(X\log D)$.

Here D is a divisor with normal crossings in the smooth complete variety X and $V = X - D$.

As an immediate consequence of (3.1)b) we have

(3.2) Proposition. The transport T satisfies

$$T := 1 + [\omega] + [\omega|\omega] + \cdots \in F^0 \cap W_0[B(E^\bullet(X \log D)) \otimes \mathbb{C}\pi_1(V,x)/J^{s+1}]\ .$$

Moreover, (3.1)a) implies that $dT = 0$ so that

$$[T] \in F^0 \cap W_0[H^0(B(E^\bullet(X\log D))) \otimes \mathbb{C}\pi_1(V,x)/J^{s+1}]\ . \ \square$$

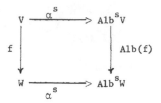

Because $dT = 0$, the value

$$<T,\gamma> = 1 + \int_\gamma \omega + \int_\gamma \omega\omega + \cdots$$

of T on a path γ depends only on the homotopy class of γ relative to its end-points. Further, (2.8) asserts that $T(\gamma) \in G_s(\mathbb{C})$ for all paths γ in V. Thus we can define a C^∞ function

$$\alpha_x^s : V \longrightarrow G_s(\mathbb{Z}) \backslash G_s(\mathbb{C})/F^0G_s \tag{3.3}$$

by taking z to the double coset of $T(\gamma)$, where γ is a path joining x to z.

Condition (3.1)b) implies that

$$\omega^{0,1} \in E^1(X \log D) \otimes F^0 \mathfrak{g}_s(\mathbb{C}) ,$$

where

$$\omega = \omega^{1,0} + \omega^{0,1}$$

is the decomposition of ω into types. From the proof of (2.8), where the derivative of the albanese mapping is computed, it follows that the mapping (3.3) is holomorphic.

As in (2.11), this construction is independent of the base point. What is not clear is that the construction of α^s is independent of the choice of the 1-form ω satisfying (3.1). This is the case. To establish this and the naturality of the construction, one needs to consider the classifying map for the unipotent variation of MHS over V whose fiber over $z \in V$ is $\mathbb{C}\pi_1(V,z)/J^{s+2}$. We leave this as a rather major exercise. Alternatively, the reader may consult [7:§5] and make the necessary translations using [4].

References

[1] Chen, K.-T., Extension of C^∞ function algebra by integrals and Malcev completion of π_1. Advances in Math, 23 (1977), 181-210.

[2] Deligne, P., Théorie d'Hodge II, Publ. Math. IHES, 40 (1971), 5-58.

[3] Deligne, P., Letter of Wojtkowiak, October 25, 1983.

[4] Hain, R., The geometry of the mixed Hodge structure on the fundamental group. To appear in Proc. of the AMS Summer Institute, Algebraic Geometry, Bowdoin College, 1985. Proc. Symp. Pure Math.

[5] Hain, R., Iterated integrals and mixed Hodge structures on homotopy groups, these proceedings.

[6] Hain, R., The de Rham homotopy theory of complex algebraic varieties II. Preprint.

[7] Hain, R., Zucker, S., Unipotent variations of mixed Hodge structure. To appear in Inventiones Math.

[8] Hwang-Ma, S.-Y., Periods of iterated integrals of holomorphic forms on a compact Riemann surface, Trans. AMS 264 (1981), 295-300.

[9] Parsin, A., A generalization of the Jacobian variety, Amer. Math. Soc. Transl. (2), 84 (1969), 187-196.

[10] Quillen, D., Rational homotopy theory, Ann. Math., 90 (1969), 205-295.

A GUIDE TO UNIPOTENT VARIATIONS OF MIXED HODGE STRUCTURE

Richard M. Hain[1]
Department of Mathematics
University of Washington
Seattle, WA 98195

Steven Zucker[2]
Department of Mathematics
The Johns Hopkins University
Baltimore, MD 21218

In [7], we gave the classification (conjectured by Deligne) of good unipotent variations of mixed Hodge structure on algebraic manifolds (and also the Kähler analogue). These are the variations whose pure weight subquotients are constant, and which behave well at infinity. The result states that a unipotent variation is the same thing as a mixed Hodge theoretic representation of the fundamental group; see Theorem (2.6) here.

Our purpose in this article is to present a more explicit treatment of the result, at least in the case of varieties X for which $H^1(X)$ is of pure weight <u>two</u>. This assumption allows us to make all constructions with global holomorphic objects, and thereby gives rise to a more constructive proof of the result. We also take the opportunity to present as an example Deligne's interpretation of the dilogarithm as part of the extension data in a unipotent variation of mixed Hodge structure (4.13); see also (3.19).

The sections of this article are:

§1. Unipotent variations
§2. Classification of unipotent variations
§3. The mixed Hodge structure on π_1
§4. Unipotent variations with trivial canonical extensions

The key point in any proof of the classification is to understand the relation between the mixed Hodge structure on π_1 and parallel transport in the flat vector bundles.

In the general case, it is natural to let classifying spaces for mixed Hodge structures and the higher Albanese manifolds of X (see [6]) play a role, but for these matters we refer to [7: §5].

---

[1]Supported in part by the National Science Foundation through grants MCS-8108814 (A04) and DMS-8401175.

[2]Supported in part by the National Science Foundation through grant DMS-8501005.

§1. Unipotent Variations

We begin by defining variations of mixed Hodge structure (MHS). Roughly speaking, a variation of MHS over the smooth complex algebraic variety X is a family of MHS's $\{V_x\}$ indexed by the points $x \in X$, subject to certain axioms; if one likes, one can regard it as a variant of a variation of Hodge structure, to which a locally constant weight filtration has been added. Additional conditions are imposed on the variation in order to control its behavior at infinity. A unipotent variation of MHS is just a variation whose monodromy representation is unipotent.

We now give the definition of a unipotent variation in detail. Let X be a smooth complex algebraic variety.

(1.1) A variation of mixed Hodge structure (MHS) $\mathcal{V}$ over X consists of:

a) a local system $\mathbb{V}_{\mathbb{Z}}$ of finitely generated abelian groups over X,

b) an increasing weight filtration $W_\bullet$ of $\mathbb{V}_{\mathbb{Q}} := \mathbb{V}_{\mathbb{Z}} \otimes \mathbb{Q}$ by local systems,

c) a decreasing Hodge filtration $\mathcal{F}^\bullet$ of the associated flat complex vector bundle $\mathcal{V} = \mathbb{V}_{\mathbb{Z}} \otimes_{\mathbb{Z}} \mathcal{O}_X$ by holomorphic sub-bundles,

d) the Hodge bundles $\{\mathcal{F}^p\}$ satisfy Griffiths' transversality. That is, with respect to the canonical flat connection ∇ on $\mathcal{V}$,

$$\nabla : \mathcal{F}^p \longrightarrow \Omega^1_X \otimes \mathcal{F}^{p-1},$$

e) for each $x \in X$, the fiber V_x over x is a mixed Hodge structure.

The variation is said to be graded-polarizable if

f) for each $k \in Z$, the filtration induced by $\mathcal{F}^\bullet$ on $Gr^W_k\mathbb{V}$ defines a polarizable variation of Hodge structure of weight k.

Such variations arise, for example, from systems of cohomology $\mathbb{V} = R^q f_* \mathbb{Z}$ associated to a topologically locally constant family of varieties $f : Z \longrightarrow X$. (See [12: §5].) Henceforth, we shall assume that all variations are graded-polarizable.

Giving the local system $\mathbb{V}_{\mathbb{Z}}$ is equivalent to specifying its monodromy representation

(1.2) $$\rho_x : \pi_1(X,x) \longrightarrow \mathrm{Aut}_{\mathbb{Z}}\, V_x \,.$$

Here we are concerned with unipotent variations of MHS, that is, variations whose monodromy representation (1.2) is unipotent. By applying the global theory of variations of Hodge structure (see [11: §7]) to (1.1, f), one sees that such variations can be characterized as follows:

(1.3) Proposition. A variation of MHS $\mathcal{V}$ is unipotent if and only if each of the variations of Hodge structure $Gr^W_k\mathbb{V}$ is constant.

The most basic examples of unipotent variations over X are those given by the MHS on its fundamental group.

(1.4) Tautological Variations of MHS. Let $x \in X$. The integral group ring of

$\pi_1(X,x)$ will be denoted by $\mathbb{Z}\pi_1(X,x)$. The homomorphism of $\pi_1(X,x)$ into the trivial group induces the algebra homomorphism

$$\varepsilon : \mathbb{Z}\pi_1(X,x) \longrightarrow \mathbb{Z}$$

on group rings. Its kernel, denoted J, is called the augmentation ideal. Fix an integer $s \geq 1$. The local system of truncated group rings

$$\mathbb{V}_{\mathbb{Z}} = \{\mathbb{Z}\pi_1(X,x)/J^{s+1}\}_{x\in X}$$

underlies a unipotent variation of MHS. The monodromy representation is induced by the inner automorphism representation

$$(1.5) \qquad \rho_x : \pi_1(X,x) \longrightarrow \mathrm{Aut}(\mathbb{Z}\pi_1(X,x)/J^{s+1})$$

$$g \longmapsto \{U \longrightarrow g^{-1}Ug\}.$$

Since ρ acts trivially on the graded quotients of the flag

$$\mathbb{C}\pi_1(X,x)/J^{s+1} \supseteq J/J^{s+1} \supseteq \ldots \supseteq J^s/J^{s+1} \supseteq 0,$$

ρ is unipotent and it induces an algebra homomorphism

$$(1.6) \qquad \bar{\rho} : \mathbb{C}\pi_1(X,x)/J^s \longrightarrow \mathrm{End}[\mathbb{C}\pi_1(X,x)/J^{s+1}].$$

Denote the lower central series of $\pi_1(X,x)$ by

$$\pi_1(X,x) = \Gamma^1 \supseteq \Gamma^2 \supseteq \ldots \quad .$$

When $\bar{\rho}$ is restricted to the Lie algebra

$$(1.7) \qquad \mathfrak{g}_s \subset \mathbb{C}\pi_1(X,x)/J^{s+1}$$

of the Malcev completion of $\pi_1(X,x)/\Gamma^{s+1}$, one obtains a Lie algebra representation

$$(1.8) \qquad d\rho_x : \mathfrak{g}_s \longrightarrow \mathrm{End}\big[\mathbb{C}\pi_1(X,x)/J^{s+1}\big],$$

which is easily seen to be the adjoint representation:

$$d\rho(A)(U) = [U,A].$$

With $\mathbb{W}_{\bullet}$ and $\mathcal{F}^{\bullet}$ appropriately defined, we get the tautological variation associated to X. For varieties satisfying $W_1H^1(X) = 0$ we will give the construction of the tautological variations in section 3.

To ensure that a variation of MHS is "reasonable", one has to impose conditions on its behavior at infinity. Let $\bar{X}$ be a smooth completion of X such that $\bar{X} - X$ is a divisor D with normal crossings in $\bar{X}$. The conditions along D are simpler for unipotent variations than in general, so we state them only in this restricted case.

(1.9) A unipotent variation of MHS $\mathcal{V}$ is good if

a) the Hodge bundles $\mathcal{F}^p$ extend over $\bar{X}$ to sub-bundles $\bar{\mathcal{F}}^p$ of the canonical extension $\bar{\mathcal{V}}$ (see, e.g., [12: p.509]) of $\mathcal{V}$ such that they induce the corresponding thing for each pure quotient $Gr_k^W \mathcal{V}$,

b) if N is a nilpotent logarithm of a local monodromy transformation about a component of D, then $NW_k \subseteq W_{k-2}$ (compare with the a priori statement:

$NW_k \subseteq W_{k-1}$).

The tautological variations associated to X are good.

§2. Classification of Unipotent Variations

Suppose that G is a group and that $\rho : G \longrightarrow \mathrm{Aut}\, V$ is a finite dimensional linear representation of G. The representation induces an algebra homomorphism

$$\bar{\rho} : \mathbb{Z}G \longrightarrow \mathrm{End}\, V$$

of the integral group ring of G. If ρ is unipotent, it follows from the Kolchin - Engel Theorem (see [10: 5.3]) and the fact that J is spanned by $\{g-1 : g \in G\}$ that $\bar{\rho}$ induces a homomorphism

$$\bar{\rho} : \mathbb{Z}G/J^n \longrightarrow \mathrm{End}\, V,$$

where $n = \dim V$.

Consequently, if $\mathcal{V}$ is a unipotent variation of MHS over the smooth variety X, then there exists $s \geq 0$ such that the monodromy representation (1.2) induces a ring homomorphism

(2.1) $$\bar{\rho}_x : \mathbb{Z}\pi_1(X,x)/J^{s+1} \longrightarrow \mathrm{End}_{\mathbb{Z}}\, V_x \,.$$

Recall that, for each point x of the variety X, each truncation $\mathbb{C}\pi_1(X,x)/J^{s+1}$ of the group ring of $\pi_1(X,x)$ carries a MHS. (See, for example, [8], [3]).

If V is a MHS, then, by standard constructions, the endomorphism ring $\mathrm{End}\, V$ has an induced MHS. Our first result says that the MHS on the truncated group ring at $x \in X$ is compatible with the MHS on the fiber V_x over x of a unipotent variation.

(2.2) Theorem. If $\mathcal{V}$ is a good unipotent variation of MHS over X, then the monodromy representation (2.1) is a morphism of MHS.

Since the Malcev Lie Algebra $\mathfrak{g}_s$ (1.7) generates $\mathbb{C}\pi_1(X,x)/J^{s+1}$ as an algebra with MHS, an equivalent formulation of (2.2) is that the monodromy representation

(2.3) $$d\rho_x : \mathfrak{g}_s \longrightarrow \mathrm{End}\, V_x$$

of a good unipotent variation is a morphism of MHS. This formulation of (2.2) is useful for checking that the theorem holds for the tautological variations.

(2.4) Example Since multiplication in the group ring $\mathbb{C}\pi_1(X,x)/J^{s+1}$ is a morphism of MHS, and since the monodromy representation (2.3) is the adjoint representation (1.8), we already know that the monodromy representations of the tautological variations over X are morphisms of MHS.

A unipotent representation ρ of $\pi_1(X,x)$ into the automorphisms of a MHS V will be called a mixed Hodge representation if the induced homomorphism

$$\bar{\rho} : \mathbb{Z}\pi_1(X,x)/J^{s+1} \longrightarrow \mathrm{End}_{\mathbb{Z}}\, V$$

is a morphism of MHS.

Theorem (2.2) enables us to define a functor as follows. Let $\underline{\underline{\mathrm{UVar}}}(X)$ be the category of good unipotent variations of MHS over X and their morphisms and let $\underline{\underline{\mathrm{HRep}}}(X,x)$ be the category of mixed Hodge representations of $\pi_1(X,x)$. Theorem (2.2) says that there is a well-defined restriction functor

$$(2.5) \qquad r_x : \underline{\underline{\mathrm{UVar}}}(X) \longrightarrow \underline{\underline{\mathrm{HRep}}}(X,x)$$
$$\mathcal{V} \longmapsto \rho_x$$

obtained by taking a variation to its monodromy representation at x.

(2.6) Theorem. The functor r_x is an equivalence of categories.

Implicit in (2.6) is the rigidity theorem: two good unipotent variations having the same fiber over x (as a MHS) and the same monodromy are isomorphic . This is a special case of the rigidity theorem for variations of MHS on algebraic varieties (cf. [12: (4.20)]).

§3. The MHS on π_1

Suppose that X is a smooth variety. Define the irregularity $q(X)$ of X to be $h^{1,0}(\bar{X})$, where $\bar{X}$ is any smooth completion of X. Varieties with irregularity zero include Zariski open subsets of simply connected projective varieties.

In this section we give an explicit construction of the MHS on $\mathbb{C}\pi_1(X,x)/J^{s+1}$ when $q(X) = 0$. However, we do not assume initially that $q = 0$.

Choose a smooth completion $\bar{X}$ of X such that $\bar{X} - X$ is a divisor D in $\bar{X}$ with normal crossings. Let

$$\Omega^p(\bar{X} \log D) = \left\{ \begin{array}{l} \text{global meromorphic p-forms on } \bar{X} \text{ that are} \\ \text{holomorphic on } X \text{ and have logarithmic poles along } D \end{array} \right\}$$

This is a finite dimensional vector space. Choose bases $w_1,\dots,w_m$ of $\Omega^1(\bar{X} \log D)$ and $z_1,\dots,z_n$ of $\Omega^2(\bar{X} \log D)$. Define complex constants a_{ij}^k by

$$(3.1) \qquad w_i \wedge w_j = \sum a_{ij}^k z_k .$$

Let $X_1,\dots,X_m$ be the dual basis of $\Omega^1(\bar{X} \log D)^*$. Denote the free associative algebra they generate by $\mathbb{C}\langle X_1,\dots,X_m\rangle$. (In other words, this is the ring of polynomials in the non-commuting indeterminates X_j .) Evaluation at 0 defines an augmentation

$$\varepsilon : \mathbb{C}\langle X_1,\dots,X_m\rangle \longrightarrow \mathbb{C} .$$

We shall denote the sth power of its augmentation ideal, $\ker \varepsilon$, by I^s. (That is, I^s is the ideal of polynomials with no terms of order less than s.) Let

(3.2) $$R_k = \sum_{i,j} a_{ij}^k [X_i, X_j] = \sum_{i,j} a_{ij}^k (X_i X_j - X_j X_i), \qquad k = 1,\dots,n,$$

and define

(3.3) $$A_s = \mathbb{C}\langle X_1,\dots,X_m\rangle / (R_1,\dots,R_n) + I^{s+1},$$

and

(3.4) $$A = A_\infty = \varprojlim A_s = \mathbb{C}\langle\langle X_1,\dots,X_m\rangle\rangle / (\overline{R_1,\dots,R_n}),$$

where $\mathbb{C}\langle\langle X_1,\dots,X_m\rangle\rangle$ denotes the ring of formal power series in the non-commuting indeterminates X_j, and the bar denotes closure in the I-adic topology. Finally, for $1 \le s \le \infty$, set

$$\omega = w_1 X_1 + \dots + w_m X_m \in \Omega^1(\overline{X} \log D) \otimes A_s .$$

Observe that the relations (3.2) guarantee that $\omega \wedge \omega = 0$. As each w_j is closed [2: (3.2.14)], ω is integrable. That is

(3.5) $$d\omega + \omega \wedge \omega = 0 .$$

Consequently, its transport

(3.6) $$T = 1 + \int\omega + \int\omega\omega + \dots = 1 + \sum \int w_i X_i + \sum \int w_i w_j X_i X_j + \dots$$

is a closed A_s-valued iterated integral; its value on a path γ depends only on the homotopy class of γ relative to its end-points (see [3]). Therefore, for each $x \in X$, T defines a function

$$T_s : \pi_1(X,x) \longrightarrow A_s \qquad 1 \le s \le \infty$$

$$\gamma \longmapsto \langle T,\gamma\rangle := 1 + \int_\gamma \omega + \int_\gamma \omega\omega + \dots .$$

Since T_s is multiplicative, it induces a homomorphism

$$T_s : \mathbb{C}\pi_1(X,x) \longrightarrow A_s \qquad 1 \le s \le \infty$$

of augmented algebras, which descends to an algebra homomorphism

(3.7) $$\Theta_s : \mathbb{C}\pi_1(X,x)/J^{s+1} \longrightarrow A_s \qquad 1 \le s < \infty .$$

Taking inverse limits we obtain

(3.8) $$\Theta = \Theta_\infty : \mathbb{C}\pi_1(X,x)^\wedge \longrightarrow A ,$$

where $\mathbb{C}\pi_1(X,x)^\wedge$ denotes the J-adic completion of $\mathbb{C}\pi_1(X,x)$.[3]

The following proposition is proved in [6: (2.4), (2.5)].

(3.9) Proposition. If $q(X) = 0$, then each Θ_s, $1 \le s \le \infty$, is an isomorphism of augmented algebras.

---

[3] If we give A the structure of a complete Hopf algebra by defining each X_j to be primitive, then Θ is a homomorphism of complete Hopf algebras.

(3.10) Remark. In general, Θ_s induces an isomorphism

$$[\mathbb{C}\pi_1(X,x)/J^{s+1}]/(F*J)^1 \cong A_s \qquad s<\infty,$$

where $(F*J)^1$ is the ideal

$$F^0 \cap J^1 + F^{-1} \cap J^2 + F^{-2} \cap J^3 + \ldots .$$

This is proved in [4: (5.2)].

For the rest of this section, we assume that $q(X) = 0$. Our next task is to describe Hodge and weight filtrations on A_s so that Θ_s is an isomorphism of bifiltered algebras.

Because $q(X) = 0$,

$$H^1(X,\mathbb{C}) \cong \Omega^1(\overline{X} \log D)$$

has a Hodge structure of pure type (1,1). Therefore each X_j has type (-1,-1). This bigrading extends naturally to a bigrading $\bigoplus_{p,q} A_s^{p,q}$ of each A_s $(s<\infty)$ as the relations (3.2) are all of type (-2,-2). Define Hodge and weight filtrations $F^\bullet$ and $W_\bullet$ on A_s in the usual way:

$$F^p A_s = \bigoplus_{u \geq p} A_s^{u,\bullet}, \qquad W_\ell A_s = \bigoplus_{p+q \leq \ell} A_s^{p,q}.$$

Note that

(3.11) $$W_{-\ell} A_s = J^k \qquad \text{if } \ell = 2k-1, 2k.$$

Define a $\mathbb{Z}$-structure on A_s by setting

(3.12) $$A_s(\mathbb{Z}) = \text{im}\,\{T_s : \mathbb{Z}\pi_1(X,x) \longrightarrow A_s\}.$$

It follows from (3.11) that the weight filtration is defined over $\mathbb{Q}$.

The next result is a straightforward consequence of the definition of the Hodge and weight filtrations on $\mathbb{C}\pi_1(X,x)/J^{s+1}$ and the proof of (3.9).

(3.13) Proposition. If $q = 0$ and $1 \leq s < \infty$, then Θ_s is an isomorphism of MHS.

(3.14) Example. When $X = \mathbb{P}^1 - \{t_0, \ldots, t_m\}$, Θ induces an isomorphism

$$\mathbb{C}\pi_1(X,x) \cong \mathbb{C}\langle\langle X_1, \ldots, X_m \rangle\rangle$$

of complete Hopf algebras (see [9: App. A]).

There is a companion description of the MHS on the Lie algebra $\mathfrak{g}_s(X,x)$ of the Malcev completion of the nilpotent quotient $\pi_1(X,x)/\Gamma^{s+1}$. Denote the free Lie algebra generated by $X_1, \ldots, X_m$ by $\mathbb{L}(X_1, \ldots, X_m)$. This can be realized as the Lie subalgebra of $\mathbb{C}\langle X_1, \ldots, X_m \rangle$ generated by $X_1, \ldots, X_m$, where $\mathbb{C}\langle X_1, \ldots, X_m \rangle$ is viewed as a Lie algebra with commutator as bracket. Set

$$\mathbb{L}^s = I^s \cap \mathbb{L}(X_1, \ldots, X_m).$$

The relations (3.2) are elements of $\mathbb{L}(X_1, \ldots, X_m)$.

Define

(3.15) $$\mathfrak{g}_s = \mathbb{L}(X_1,\ldots,X_m)/(R_1,\ldots,R_n) + \mathbb{L}^{s+1}$$

and

(3.16) $$\mathfrak{g} = \varprojlim \mathfrak{g}_s .$$

Using the Poincaré-Birkhoff-Witt theorem, one can show that the natural mapping $\mathfrak{g}_s \longrightarrow A_s$ is injective.

(3.17) Proposition. Under the isomorphism Θ, $\mathfrak{g}$ corresponds to the Malcev Lie algebra associated with $\pi_1(X,x)$. Consequently, when $s<\infty$, $\mathfrak{g}_s$ is the Lie algebra of the Malcev completion of $\pi_1(X,x)/\Gamma^{s+1}$.

Proof (Sketch). The Malcev Lie algebra associated with $\pi_1(X,x)$ is, by definition [9: Appendix A3], the set of primitive elements of the complete Hopf algebra $\mathbb{C}\pi_1(X,x)^\wedge$. The result follows from the fact that the set of primitive elements of $\mathbb{C}\langle\langle X_1,\ldots,X_m\rangle\rangle$ is the completion of $\mathbb{L}(X_1,\ldots,X_m)$ [9: A2.11].

(3.18) Corollary. The Malcev completion of $\pi_1(X,x)/\Gamma^{s+1}$ is isomorphic to

$$G_s(\mathbb{C}) = \{\exp U \in A_s : U \in \mathfrak{g}_s\} \subseteq A_s$$

via Θ_s.

(3.19) Example. If $X = \mathbb{P}^1-\{0,1,\infty\}$ and $s = 2$, then

$$\mathfrak{g}_2 = \mathbb{L}(X_0,X_1)/\mathbb{L}^3 \cong \left\{\begin{pmatrix} 0 & a_1 & a_3 \\ 0 & 0 & a_2 \\ 0 & 0 & 0 \end{pmatrix} : a_i \in \mathbb{C}\right\}$$

and

$$G_2 \cong \left\{\begin{pmatrix} 1 & b_1 & b_3 \\ 0 & 1 & b_2 \\ 0 & 0 & 1 \end{pmatrix} : b_i \in \mathbb{C}\right\}$$

is the Heisenberg group. The map Θ_2 induces a homomorphism

$$\pi_1(\mathbb{P}^1-\{0,1,\infty\},x) \longrightarrow G_2$$

$$\gamma \longmapsto \begin{pmatrix} 1 & \int_\gamma \frac{dx}{1-x} & \int_\gamma \frac{dx}{1-x}\frac{dx}{x} \\ 0 & 1 & \int_\gamma \frac{dx}{x} \\ 0 & 0 & 1 \end{pmatrix}$$

Note that $\int_\gamma \frac{dx}{1-x}\frac{dx}{x}$ is essentially a dilogarithm as

$$L_2(z) = \frac{z}{1^2} + \frac{z^2}{2^2} + \frac{z^3}{3^2} + \dots = \int_0^z \frac{dx}{1-x}\frac{dx}{x} .$$

(3.20) Realization of the tautological variation. The s-th tautological variation of a variety X with $q=0$ can be described explicitly as follows. The canonical extension of the variation is the trivial bundle

$$A_s \times \overline{X} \longrightarrow \overline{X}$$

and the extended Hodge bundle $\overline{\mathcal{F}}^p$ and the weight bundle $\overline{\mathcal{W}}_\ell$ are given by

$$F^pA_s \times \overline{X} \longrightarrow \overline{X} , \qquad W_\ell A_s \times \overline{X} \longrightarrow \overline{X} ,$$

respectively. The canonical connection on $\overline{\mathcal{V}}$ is given by

$$\nabla s = ds - s\omega ,$$

where $s \in \mathcal{O}(\overline{\mathcal{V}})$,

$$\omega = w_1X_1 + \dots + w_mX_m \in \Omega^1(\overline{X} \log D) \otimes \mathfrak{g}_{s-1}$$

and $\mathfrak{g}_{s-1}$ is viewed as a Lie subalgebra of $\operatorname{End} A_s$ via the adjoint action:

$$\begin{aligned} \mathrm{ad} : \mathfrak{g}_{s-1} &\longrightarrow \operatorname{End} A_s \\ U &\longmapsto \{Y \longmapsto [Y,U]\} . \end{aligned}$$

Since

$$X_j \in (F^{-1} \cap W_{-2})\operatorname{End} A_s ,$$

it follows that:

a) the weight filtration is flat and the induced connection is trivial on $Gr^W \overline{\mathcal{V}}$,

b) $\nabla\mathcal{F}^p \subseteq \mathcal{F}^{p-1}$,

c) the variation is good.

The $\mathbb{Z}$ structure given by (3.12) is easily seen to be flat.

§4. Unipotent Variations with Trivial Canonical Extension

Throughout this section, X will be a smooth algebraic variety and $\overline{X}$ a smooth completion of X such that $\overline{X}-X$ is a divisor D in $\overline{X}$ with normal crossings.

In this section we give an explicit description of good unipotent variations over X with holomorphically trivial canonical extension, and prove (2.2) for such variations. When $q(X) = 0$, all unipotent variations over X have trivial canonical extension, and we give a constructive proof of the equivalence of categories (2.6).

We begin by recalling the criterion for the triviality of the canonical extension of a flat bundle with unipotent monodromy.

(4.1) Lemma [4: (6.4)]. A flat bundle $E \to X$ with unipotent monodromy has trivial canonical extension if and only if the kernel of the induced representation

$$\bar{\rho} : \mathbb{C}\pi_1(X,x)/J^{s+1} \longrightarrow \mathrm{End}\, E_x$$

contains the ideal

$$(F*J)^1 = F^0 \cap J^1 + F^{-1} \cap J^2 + \ldots \quad .$$

If $q(X) = 0$, then it follows from (3.13) that $(F*J)^1 = 0$.

(4.2) Corollary. If $q(X) = 0$, then every flat bundle over X with unipotent monodromy has trivial canonical extension.

By appealing to (2.2), we can give more examples of unipotent variations with trivial canonical extension.

(4.3) Corollary. If $\mathcal{V}$ is a good unipotent variation over X satisfying

(4.4) $$(F^0 \cap W_{-1})\mathrm{End}\, V_x = 0 \, ,$$

then $\mathcal{V}$ has trivial canonical extension.

Proof. Since $J \subseteq W_{-1}\mathbb{C}\pi_1(X,x)/J^{s+1}$, it follows from (2.2) and (4.4) that the kernel of the monodromy representation $\bar{\rho}$ contains $F^0 \cap J$. The result now follows from (4.1) and the fact that $(F*J)^1$ is the ideal of $\mathbb{C}\pi_1(X,x)/J^{s+1}$ generated by $F^0 \cap J$.

Since a MHS that is an iterated extension of Tate structures $\mathbb{Z}(m)$ satisfies (4.4), we have:

(4.5) Corollary. If $\mathcal{V}$ is a good unipotent variation whose weight graded quotients are direct sums of Tate structures, then $\mathcal{V}$ has trivial canonical extension.

Unipotent variations with trivial canonical extension enjoy the property that their extended Hodge and weight bundles are also trivial. For example, we have already seen in (3.20) that this is true for the tautological variations associated to a variety with $q=0$.

(4.6) Proposition. If $\mathcal{V}$ is a good unipotent variation with trivial canonical extension, then the extended Hodge and weight bundles are trivial. That is, if $\bar{\mathcal{V}} = V \times \bar{X} \to \bar{X}$, then

$$\bar{\mathcal{F}}^p = F^p V \times \bar{X} \to \bar{X} \, , \quad \bar{\mathcal{W}}_\ell = W_\ell V \times \bar{X} \to \bar{X}$$

and the embeddings of $\bar{\mathcal{F}}^p$ and $\bar{\mathcal{W}}_\ell$ into $\bar{\mathcal{V}}$ are induced by the inclusions $F^p V \to V$, $W_\ell V \to V$.

Proof. Each $\mathcal{W}_\ell$ is a flat subbundle of $\mathcal{V}$. Consequently, the vanishing of the monodromy of $\mathcal{V}$ on $(F{*}J)^1$ implies the same for $\mathcal{W}_\ell$ which, by (4.1), implies that $\overline{\mathcal{W}}_\ell$ is trivial. Suppose that

$$\overline{\mathcal{V}} = V{\times}\overline{X} \to \overline{X}\ , \text{ and } \overline{\mathcal{W}}_\ell = W_\ell V{\times}\overline{X} \to \overline{X}\ .$$

The inclusion $\overline{\mathcal{W}}_\ell \to \overline{\mathcal{V}}$ corresponds to a holomorphic function $\overline{X} \to \mathrm{Hom}(W_\ell V, V)$, a function from a complete variety to affine space, and hence constant.

We cannot use the same argument to prove the triviality of $\mathcal{F}^p$, as the connection does not preserve the Hodge filtration. Denote by $W_{-1}\,\mathrm{Aut}\,V$, $F^0W_{-1}\,\mathrm{Aut}\,V$ the simply connected nilpotent subgroups of $\mathrm{Aut}\,V$ corresponding to the nilpotent Lie algebras $W_{-1}\,\mathrm{End}\,V$, $(F^0 \cap W_{-1})\mathrm{End}\,V$, respectively. The embeddings $\mathcal{F}^p \to \overline{\mathcal{V}}$ are given by a holomorphic mapping

$$\overline{X} \longrightarrow W_{-1}\,\mathrm{Aut}\,V\ /\ F^0W_{-1}\,\mathrm{Aut}\,V\ ,$$

as the right-hand side parametrizes liftings of $\mathcal{F}^{\bullet}\mathrm{Gr}^W V$. The triviality of the Hodge filtration now follows from the following fact:

(4.7) Lemma. If H is a subgroup of the simply connected, complex, nilpotent Lie group G, then G/H is biholomorphic to $\mathbb{C}^N$.

Proof. Denote the Lie algebras of H and G by $\mathfrak{h}$ and $\mathfrak{g}$, respectively. Since $\mathfrak{g}$ is nilpotent, we can find a chain of subalgebras

$$\mathfrak{h} = \mathfrak{h}_0 \subset \mathfrak{h}_1 \subset \ldots \subset \mathfrak{h}_N = \mathfrak{g}$$

with

$$\dim(\mathfrak{h}_j/\mathfrak{h}_{j-1}) = 1 \qquad j=1,\ldots,N.$$

Let $X_j \in \mathfrak{h}_j - \mathfrak{h}_{j-1}$. Then the holomorphic function $\mathbb{C}^N \to G$ defined by

$$(x_1,\ldots,x_N) \longmapsto \exp(x_1X_1)\cdot\exp(x_2X_2)\cdot\ldots\cdot\exp(x_NX_N)$$

induces a holomorphic mapping $\mathbb{C}^N \to G/H$. By induction on N, the latter is easily seen to be a biholomorphism.

We can now give a direct proof of (2.2) for unipotent variations with trivial canonical extension. (We admit that for variations satisfying $(F^0 \cap W_{-1})\mathrm{End}\,V_x = 0$, this argument is circular, as we have made use of (2.2) in (4.3) to prove that such variations have trivial canonical extension.)

(4.8) Proposition. The monodromy representation

(4.9) $$\bar{\rho} : \mathbb{C}\pi_1(X,x)/J^{s+1} \longrightarrow \mathrm{End}\,V_x$$

of a good unipotent variation with trivial canonical extension is a morphism of MHS.

Proof. Let $\mathcal{V} \to X$ be a good unipotent variation with trivial canonical extension. Since its monodromy representation (4.9) is defined over $\mathbb{Z}$, to prove that $\bar{\rho}$ is a morphism of MHS, it suffices to prove that it preserves the Hodge and weight filtrations; that is,

(4.10) $$\bar{\rho} \in (F^0 \cap W_0)[(\mathbb{C}\pi_1(X,x)/J^{s+1})^* \otimes \operatorname{End} V_x].$$

Because $\bar{\mathcal{V}}$ is trivial and because the canonical connection ∇ has regular singularities along D, ∇ is given by

$$\nabla s = ds - s\omega \qquad s \in \mathcal{O}_{\bar{X}} \otimes V_x ,$$

where

$$\omega \in \Omega^1(\bar{X} \log D) \otimes \operatorname{End} V_x$$

As the Hodge and weight bundles are trivial (4.6), the fact that $\mathcal{V}$ is a good variation implies that

$$\omega \in (F^0 \cap W_0)[\Omega^1(\bar{X} \log D) \otimes \operatorname{End} V_x].$$

(Here the weights on $\Omega^1(\bar{X} \log D)$ have been shifted, as in $H^1(X,\mathbb{C})$.) The associated transport

$$T = 1 + \int\omega + \int\omega\omega + \ldots$$

determines an element

$$[T] \in (F^0 \cap W_0)[(\mathbb{C}\pi_1(X,x)/J^{s+1})^* \otimes \operatorname{End} V_x] ,$$

which can be shown to be $\bar{\rho}$ [3: (2.6),(5.1)].

We conclude this section with a direct construction of a good variation with trivial extension from a mixed Hodge representation of $\pi_1(X,x)$ that vanishes on $(F*J)^1$. This, together with (4.8), completes the proof of (2.6) for varieties with $q = 0$.

(4.11) Construction. Suppose that V is a MHS and that

$$\rho: \pi_1(X,x) \longrightarrow \operatorname{Aut}_{\mathbb{Z}} V$$

is a unipotent representation that induces a morphism of MHS

$$\bar{\rho}: \mathbb{C}\pi_1(X,x)/J^{s+1} \longrightarrow \operatorname{End} V$$

satisfying

$$\ker \bar{\rho} \supseteq (F*J)^1.$$

Define the bundles $\bar{\mathcal{V}}, \mathcal{F}^p, \bar{\mathcal{W}}_\ell$ to be

$$\begin{array}{ccc} V\times\bar{X} & F^pV\times\bar{X} & W_\ell V\times\bar{X} \\ \downarrow & \downarrow & \downarrow \\ \bar{X} & X & \bar{X} \end{array}$$

respectively. Choose a basis $w_1,\ldots,w_m$ of $\Omega^1(\bar{X} \log D)$ and a dual basis $X_1,\ldots,X_m$ of its dual. Recall from (3.10) that

$$[\mathbb{C}\pi_1(X,x)/J^{s+1}]/(F*J)^1 \cong \mathbb{C}\langle X_1,\ldots,X_m\rangle/(R_1,\ldots,R_m) + I^{s+1}.$$

The universal connection form

$$\tilde{\omega} = \sum w_j X_j$$

is integrable and lies in

$$(F^0 \cap W_0)(\Omega^1(\overline{X} \log D) \otimes [\mathbb{C}\pi_1(X,x)/J^{s+1}]/(F*J)^1) .$$

Take

$$\omega = (\mathrm{id} \otimes \overline{\rho})\tilde{\omega} .$$

This is an integrable 1-form. Since $\overline{\rho}$ preserves the filtrations,

(4.12) $\quad \omega \in (F^0 \cap W^0)(\Omega^1(\overline{X} \log D) \otimes \mathrm{End}\, V).$

Define a connection ∇ on $\overline{\mathcal{V}}$ by

$$\nabla s = ds - s\omega \qquad s \in \mathcal{O}_X \otimes V.$$

Since ω is integrable, this defines a flat connection on $\mathcal{V}$ whose canonical extension is $\overline{\mathcal{V}}$. Define a $\mathbb{Z}$-structure on the fiber V_y over $y \in X$ by parallel transporting $V_{\mathbb{Z}}$ from to V_x to V_y. This gives a rational structure to each fiber. Because of (4.12), the weight filtration is flat. This, together with the fact that the weight filtration on V_x is defined over $\mathbb{Q}$, implies the same for each fiber. It is an exercise to check that (4.12) implies that $\overline{\mathcal{V}}$ is a good variation of MHS.

(4.13) <u>Example</u>. In this example we construct the variation of MHS over $X = \mathbb{P}^1 - \{0,1,\infty\}$ with graded quotients $\mathbb{Z}(0)$, $\mathbb{Z}(1)$, $\mathbb{Z}(2)$ associated to the dilogarithm $L_2(x)$ (3.19). This example is due to P. Deligne.

Fix $x \in X$. Let e_0, e_1, e_2 be the standard basis of $V_{\mathbb{C}} = \mathbb{C}^3$. Define Hodge and weight filtrations on $V_{\mathbb{C}}$ by letting

$$F^{-2} = \mathbb{C}^3, \quad F^{-1} = \mathrm{span}\{e_0, e_1\}, \quad F^0 = \mathrm{span}\{e_0\},$$

$$W_{-4} = \mathrm{span}\{e_2\}, \quad W_{-2} = \mathrm{span}\{e_1, e_2\}, \quad W_0 = \mathbb{C}^3.$$

Choose branches of $\log z$, $\log(1-z)$ and $L_2(z)$ in a neighbourhood of x. Define $V_{\mathbb{Z}}$ to be the $\mathbb{Z}$-span of the vectors

$$v_0 = e_0 - \log(1-x)e_1 + L_2(x)e_2 ,$$

$$v_1 = 2\pi i\, e_1 + 2\pi i \log x\, e_2 ,$$

$$v_2 = (2\pi i)^2 e_2 .$$

Defining the weight filtration on $V_{\mathbb{Q}}$ in the obvious way, we obtain a MHS V_x whose graded quotients are $\mathbb{Z}(0)$, $\mathbb{Z}(1)$ and $\mathbb{Z}(2)$. Set

$$w_0 = \frac{dz}{z}, \quad \text{and} \quad w_1 = \frac{dz}{1-z} .$$

Define a homomorphism

$$\rho : \pi_1(X,x) \longrightarrow \mathrm{Aut}\, V_{\mathbb{C}}$$

by $\gamma \longmapsto T(\gamma)$, where T is the $GL(3)$ valued iterated integral

$$\begin{pmatrix} 1 & \int w_1 & \int w_1 w_0 \\ 0 & 1 & \int w_0 \\ 0 & 0 & 1 \end{pmatrix}$$

Note that GL(3) acts on the right on $\mathbb{C}^3 = V_{\mathbb{C}}$.

We have to check that ρ is defined over $\mathbb{Z}$. Choose loops σ_0, σ_1 in X based at x such that

$$\int_{\sigma_k} w_j = 2\pi i\ \delta_{jk} .$$

Because

$$\int_{\sigma_0} w_1 w_0 = - \int_{\sigma_0} [\log(1-z) - \log(1-x)] \frac{dz}{z} = 2\pi i \log(1-x)$$

and

$$\int_{\sigma_1} w_1 w_0 = \int_{\sigma_1} w_1 \int_{\sigma_1} w_0 - \int_{\sigma_1} w_0 w_1 \qquad [3: (2.11)]$$

$$= -\int_{\sigma_1} (\log z - \log x) \frac{dz}{1-z}$$

$$= 2\pi i \log x ,$$

the matrices of $T(\sigma_0)$ and $T(\sigma_1)$, with respect to the basis v_0, v_1, v_2, are

$$\begin{pmatrix} 1 & 0 & 0 \\ 0 & 1 & 1 \\ 0 & 0 & 1 \end{pmatrix} , \quad \begin{pmatrix} 1 & 1 & 0 \\ 0 & 1 & 0 \\ 0 & 0 & 1 \end{pmatrix}$$

respectively. That is, ρ is defined over $\mathbb{Z}$.

Thus, by the construction (4.11), there is a good unipotent variation over X with fiber V_x over x. One can check, using the construction, that the fiber V_y over y is $V_{\mathbb{C}}$ with the lattice spanned by the expected:

$$v_0(y) = e_0 - \log(1-y)\, e_1 + L_2(y)\, e_2 ,$$

$$v_1(y) = 2\pi i\, e_1 + 2\pi i \log y\, e_2 ,$$

$$v_2(y) = (2\pi i)^2\, e_2 .$$

References

[1] Chen, K.-T., Iterated path integrals. Bull. Amer. Math. Soc. 83 (1977), 831-879.

[2] Deligne, P., Théorie de Hodge, II. Publ. Math. IHES 40 (1971), 5-57.

[3] Hain, R., The geometry of the mixed Hodge structure on the fundamental group. To appear in Proc. of the AMS Summer Institute, Algebraic Geometry, Bowdoin College, 1985. Proc. Symp. Pure Math.

[4] Hain, R., On a generalization of Hilbert's 21st problem. To appear in Ann. scient. Ec. Norm. Sup.

[5] Hain, R., Iterated integrals and mixed Hodge structures on homotopy groups, these proceedings

[6] Hain, R., Higher albanese manifolds, these proceedings

[7] Hain, R., Zucker, S., Unipotent variations of mixed Hodge structure. To appear in Inventiones Math.

[8] Morgan, J., The algebraic topology of smooth algebraic varieties. Publ. Math. IHES 48 (1978), 137-204.

[9] Quillen, D., Rational homotopy theory. Ann. Math. 90 (1969), 205-295.

[10] Serre, J.-P., Lie Algebras and Lie Groups, Benjamin/Cummings, London, 1965.

[11] Schmid, W., Variation of Hodge structure: the singularities of the period mapping. Invent. Math. 22 (1973), 211-319.

[12] Steenbrink, J., Zucker, S., Variation of mixed Hodge structure, I. Invent. Math. 80 (1985), 489-542.

TRUNCATIONS OF MIXED HODGE COMPLEXES

Richard M. Hain[1]
Department of Mathematics
University of Washington
Seattle, WA 98195 USA

Steven Zucker[2]
Department of Mathematics
The Johns Hopkins University
Baltimore, MD 21218 USA

In this article, we show that the truncations of a mixed Hodge complex are again mixed Hodge complexes. As the main application of this result, we give the construction of a Hodge complex for the intersection homology of a projective variety with isolated singularities.

We thank P. Deligne for helpful conversations.

§1. Preliminaries on truncation functors.

Let $K^\bullet$ be a complex in an abelian category. We recall the two methods of filtering $K^\bullet$ by canonical truncation:

$$(1)\qquad (\tau_{\le n}K)^i = \begin{cases} K^i & \text{if } i < n \\ \ker d \cap K^n & \text{if } i = n \\ 0 & \text{if } i > n; \end{cases}$$

$$(2)\qquad (\tilde{\tau}_{\le n}K)^i = \begin{cases} K^i & \text{if } i \le n \\ dK^n & \text{if } i = n+1 \\ 0 & \text{if } i > n+1 \end{cases}$$

Clearly, the filtrations τ and $\tilde{\tau}$ are functorial in $K^\bullet$. One also puts

$$(3)\qquad \tau^{>n}K^\bullet = K^\bullet/\tau_{\le n}K^\bullet \,, \quad \tilde{\tau}^{>n}K^\bullet = K^\bullet/\tilde{\tau}_{\le n}K^\bullet \,.$$

The following facts are elementary:

[1]Supported in part by the National Science Foundation through grants MCS-8108814 (A04) and DMS-8401175.

[2]Supported in part by the National Science Foundation through grant DMS-8501005.

Proposition 1: i) $\tilde{\tau}_{\leq n}K^{\bullet}/\tau_{\leq n}K^{\bullet}$ is the acyclic complex

$$0 \longrightarrow K^n/\ker d \longrightarrow dK^n \longrightarrow 0 ,$$

ii) There are quasi-isomorphisms

$$Gr^n_{\tau}K^{\bullet} \tilde{\to} Gr^n_{\tilde{\tau}}K^{\bullet} \tilde{\to} H^n(K^{\bullet})[-n] ,$$

iii) $H^{\bullet}(\tau_{\leq n}K^{\bullet}) \simeq H^{\bullet}(\tilde{\tau}_{\leq n}K^{\bullet}) \simeq \tau_{\leq n}H^{\bullet}(K^{\bullet})$,

iv) $H^{\bullet}(\tau^{>n}K^{\bullet}) \simeq H^{\bullet}(\tilde{\tau}^{>n}K^{\bullet}) \simeq \tau^{>n}H^{\bullet}(K^{\bullet})$.

Assume now that $K^{\bullet}$ comes with an increasing filtration W. Of course, there are induced filtrations on the truncated complexes (1) - (3).

We say that, for a filtered complex $K^{\bullet}$, d is strictly compatible with W at degree $n+1$ if for all k

(4) $$dK^n \cap W_kK^{n+1} = d(W_kK^n)$$

(cf. [2: (1.1.5)]).

Proposition 2: i) $\tau_{\leq n}(W_kK^{\bullet}) = W_k(\tau_{\leq n}K^{\bullet})$, and likewise for $\tau^{>n}$;

ii) The following are equivalent formulations of (4):

a) for all k, $\tilde{\tau}_{\leq n}(W_kK^{\bullet}) = W_k(\tilde{\tau}_{\leq n}K^{\bullet})$, or the analogous assertion for $\tilde{\tau}^{>n}$.

b) for all k, $Gr^W_k(\tau_{\leq n}K^{\bullet}) \simeq \tau_{\leq n}(Gr^W_kK^{\bullet})$, or the analogous assertion for $\tau^{>n}$, $\tilde{\tau}_{\leq n}$, $\tilde{\tau}^{>n}$

c) the quasi-isomorphism $\tau_{\leq n}K^{\bullet} \longrightarrow \tilde{\tau}_{\leq n}K^{\bullet}$ is a filtered quasi-isomorphism with respect to W.

Remark: The above applies to decreasing filtrations, by the usual trick of reversing the degrees.

§2. Truncations of mixed Hodge complexes.

We will give the proof of the following surprising fact:

Theorem: Suppose that $(K^{\bullet}_{\mathbb{Q}},W_{\mathbb{Q}})$, $(K^{\bullet},W,F)$ comprise the data of a mixed Hodge complex defined over $\mathbb{Q}$. Then

$$(\tau_{\leq n}K^{\bullet}_{\mathbb{Q}},W_{\mathbb{Q}}),\ (\tau_{\leq n}K^{\bullet},W,F)$$

is also a mixed Hodge complex defined over $\mathbb{Q}$; the same holds for $\tilde{\tau}_{\leq n}$, $\tau^{>n}$ and $\tilde{\tau}^{>n}$.

Remark: One can replace $\mathbb{Q}$ by other sub-rings of $\mathbb{R}$.

The proof of the theorem breaks up into a sequence of propositions, which we state and prove only for $\tilde{\tau}^{>n}$ (the other cases are similar), in accordance with our applications in §3.

Proposition 3: Let $(K^\bullet,W) \longrightarrow (L^\bullet,W)$ be a filtered quasi-isomorphism of complexes with finite filtrations. Then

$$(\tilde{\tau}^{>n}K^\bullet,W) \longrightarrow (\tilde{\tau}^{>n}L^\bullet,W)$$

is also a filtered quasi-isomorphism.

Proposition 4: Suppose that the spectral sequence for W on $K^\bullet$ degenerates at E_r. Then the spectral sequence for W on $\tilde{\tau}^{>n}K$ also degenerates at E_r.

Proposition 5: If d is strictly compatible with F on $Gr_k^W K^\bullet$, then the same is true on $Gr_k^W(\tilde{\tau}^{>n}K^\bullet)$.

Proposition 6: Under the hypotheses of the theorem, F induces a Hodge structure of weight $i+k$ on $H^i(Gr_k^W(\tilde{\tau}^{>n}K^\bullet))$ for all i.

The common theme in the proofs of the four propositions above is making use of the fact that, by definition ((1)), $\tilde{\tau}^{>n}K$ differs from $K^\bullet$ or the zero complex only in degree $n+1$.

Proof of Proposition 3: Given the filtered complex $(K^\bullet,W)$, we consider the short exact sequence

$$0 \longrightarrow \tilde{\tau}_{\leq n}K^\bullet \longrightarrow K^\bullet \longrightarrow \tilde{\tau}^{>n}K^\bullet \longrightarrow 0.$$

For the induced filtrations, the formation of Gr_k^W is exact, so we get

$$0 \longrightarrow Gr_k^W(\tilde{\tau}_{\leq n}K^\bullet) \longrightarrow Gr_k^W K^\bullet \longrightarrow Gr_k^W(\tilde{\tau}^{>n}K^\bullet) \longrightarrow 0.$$

Upon taking cohomology, we get:

$$\text{(5)}\qquad \begin{aligned} &\text{i)}\ \ H^i(Gr_k^W(\tilde{\tau}^{>n}K^\bullet)) = 0 \qquad \text{if } i < n+1,\\ &\text{ii)}\ \ H^i(Gr_k^W K^\bullet) \simeq H^i(Gr_k^W(\tilde{\tau}^{>n}K^\bullet)) \quad \text{if } i > n+1,\\ &\text{iii)}\ \ H^{n+1}(Gr_k^W K^\bullet) \twoheadrightarrow H^{n+1}(Gr_k^W(\tilde{\tau}^{>n}K^\bullet)). \end{aligned}$$

The assertions analogous to (5) are, of course, valid for $L^\bullet$, and we are assuming that for all i and k

$$\text{(6)}\qquad H^i(Gr_k^W K^\bullet) \xrightarrow{\sim} H^i(Gr_k^W L^\bullet).$$

From (6), we get immediately that

$$\text{(7)}\qquad H^i(Gr_k^W(\tilde{\tau}^{>n}K^\bullet)) \longrightarrow H^i(Gr_k^W(\tilde{\tau}^{>n}L^\bullet))$$

is an isomorphism for $i \neq n+1$. At first glance, (7) is only a surjection when $i = n+1$, but we can appeal to the fact that it abuts to the isomorphism (see Prop. 1 (iv))

$$(8)\qquad H^{n+1}(K^\bullet) \simeq H^{n+1}(\tilde{\tau}^{>n}K^\bullet) \longrightarrow H^{n+1}(\tilde{\tau}^{>n}L^\bullet) \simeq H^{n+1}(L^\bullet).$$

We write

$$(9)\qquad \bigoplus_k H^i(Gr_k^W(\tilde{\tau}^{>n}K^\bullet)) = E_1^i(\tilde{\tau}^{>n}K^\bullet,W), \text{ etc.}$$

The spectral sequence for W gives recursively

$$(10)\qquad \begin{array}{ccc} 0 & & 0 \\ \downarrow d_r & & \downarrow d_r \\ E_r^{n+1}(\tilde{\tau}^{>n}K^\bullet,W) & \xrightarrow{\pi_r} & E_r^{n+1}(\tilde{\tau}^{>n}L^\bullet,W) \\ \downarrow d_r & & \downarrow d_r \\ E_r^{n+2}(\tilde{\tau}^{>n}K^\bullet,W) & \xrightarrow{\sim} & E_r^{n+2}(\tilde{\tau}^{>n}L^\bullet,W) \end{array}$$

If π_1 had a non-zero kernel, the same would be true in the limit, a contradiction. Thus, (7) is an isomorphism for $i = n+1$ too, as desired.

<u>Proof of Proposition 4</u>: The argument is similar to the one involving (10). Consider

$$(11)\qquad \begin{array}{ccc} E_r^{i-1}(K^\bullet,W) & \longrightarrow & E_r^{i-1}(\tilde{\tau}^{>n}K^\bullet,W) \\ \downarrow d_r & & \downarrow d_r \\ E_r^{i}(K^\bullet,W) & \longrightarrow & E_r^{i}(\tilde{\tau}^{>n}K^\bullet,W) \\ \downarrow d_r & & \downarrow d_r \\ E_r^{i+1}(K^\bullet,W) & \longrightarrow & E_r^{i+1}(\tilde{\tau}^{>n}K^\bullet,W) \end{array}$$

We know the behavior of

$$(12)\qquad E_1^i(K^\bullet,W) \longrightarrow E_1^i(\tilde{\tau}^{>n}K^\bullet,W)$$

from (5). The mapping (12) is an isomorphism for $i > n+1$, a surjection for $i = n+1$, and zero otherwise. Via (11), we see recursively that this persists for all E_r. The desired conclusion follows.

<u>Proof of Proposition 5</u>: Because $\tilde{\tau}^{>n}K^\bullet$ satisfies (3), the strictness of d with respect to F is immediate, and is left for the reader to check.

<u>Proof of Proposition 6</u>: The only non-trivial issue is to show that F induces a Hodge structure of weight $n+1+k$ on

$$H^{n+1}(Gr_k^W(\tilde{\tau}^{>n}K^\bullet)).$$

For this, consider the diagram

(13)
$$\begin{array}{ccccc}
H^{n+1}(Gr_k^W F^p K^\cdot) & H^n(Gr_{k+1}^W K^\cdot) & & 0 & H^{n+1}(Gr_k^W(\tilde{\tau}^{>n} F^p K^\cdot)) \\
\searrow & \downarrow d_1 & & \downarrow \tilde{d}_1 & \swarrow \\
& H^{n+1}(Gr_k^W K^\cdot) & \longrightarrow & H^{n+1}(Gr_k^W(\tilde{\tau}^{>n} K^\cdot)) & \\
& \downarrow d_1 & & \downarrow \tilde{d}_1 & \\
& H^{n+2}(Gr_{k-1}^W K^\cdot) & \xrightarrow{\sim} & H^{n+2}(Gr_{k-1}^W(\tilde{\tau}^{>n} K^\cdot)) &
\end{array}$$

(with a curved arrow along the top from $H^{n+1}(Gr_k^W F^p K^\cdot)$ to $H^{n+1}(Gr_k^W(\tilde{\tau}^{>n} F^p K^\cdot))$)

That F^p on cohomology is induced by the injective images from the corners of the diagram follows from the strictness of F (see Prop. 5 and Prop. 2 (ii,a)). By Prop. 4, the spectral sequence of W on $\tilde{\tau}^{>n}K^\cdot$ degenerates at E_2; moreover, it abuts to $H^{n+1}(K^\cdot)$ in degree $n+1$. This gives

(14) $$H^{n+1}(Gr_k^W(\tilde{\tau}^{>n}K^\cdot)) \cong H^{n+1}(Gr_k^W K^\cdot)/\text{im } d_1 .$$

Now, by assumption, F induces a Hodge structure of weight $n+1+k$ on the right-hand side of (14). It remains to verify that the filtrations F coincide under the isomorphism, but this is clear from the top line of the diagram (13).

§3. Applications.

One direct application of the theorem from §2 is to reduced bar constructions in [5: §3]. There, we needed to know that if $K^\cdot$ is a mixed Hodge complex that is zero in negative degrees, then

$$K^+/dK^0 = \tilde{\tau}^{>0}K^\cdot$$

is also a mixed Hodge complex.

There is a second, more interesting, application to the intersection cohomology of varieties with isolated singularities. Let X be an m-dimensional projective variety with 0-dimensional singular locus Σ. One has the well-known formulas:

(15) $$IH^i(X) \cong \begin{cases} H^i(X-\Sigma) & \text{if } i < m, \\ H^i(X) & \text{if } i > m, \\ \text{Im}\{H^m(X) \to H^m(X-\Sigma)\} & \text{if } i = m. \end{cases}$$

With the intersection cohomology groups expressed in terms of ordinary cohomology, these groups inherit mixed Hodge structures from the construction in [2]. It is known that these Hodge structures are, in fact, pure [6: §3], [7: (1.14)].

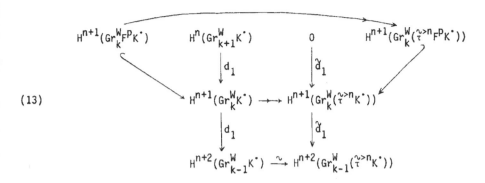

With the help of the result from §2, we can construct at least a mixed Hodge complex (not a cohomological one) for the intersection cohomology. There is a triangle

$$\begin{array}{ccc} IH^\bullet(X) & \longrightarrow & H^\bullet(X-\Sigma) \\ & {\scriptstyle +1}\nwarrow \quad \swarrow & \\ & \tau^{>m-1}H^\bullet(U-\Sigma), & \end{array}$$

where U is a small neighborhood of Σ in X. To take advantage of this, we need compatible mixed Hodge complexes for the cohomology of $X-\Sigma$ and $U-\Sigma$, which we denote respectively $K^\bullet$ and $L^\bullet$. (Such do exist!) We then have a morphism ϕ of mixed Hodge complexes, defined by the composite

$$K^\bullet \longrightarrow L^\bullet \longrightarrow \tilde{\tau}^{>m-1}L^\bullet.$$

Let $C^\bullet = C_M(\phi)$ be the mixed cone [4: II(1.2)] (see also [3: (2.3)]). It is a mixed Hodge complex that induces on cohomology the (pure) mixed Hodge structures on (15). Of course, $C^\bullet$ has a non-trivial weight filtration, but all of its cohomology comes from that of $Gr_0^W C^\bullet$. Since we really don't care about W, we should just disregard it and consider $C^\bullet$ as a Hodge complex of weight 0! Once we do that, we see that we can replace $K^\bullet$ and $L^\bullet$ by F-filtered quasi-isomorphic complexes (that have only one filtration).

Good choices for $K^\bullet$ and $L^\bullet$ are as follows. Let $\tilde{X}$ be a desingularization of X, in which Σ has been replaced by D, a divisor with normal crossings. Then take $K^\bullet$ to be the Dolbeault or Cech complex for $\Omega^\bullet_{\tilde{X}}(\log D)$; for $L^\bullet$, do likewise for

$$\frac{\Omega^\bullet_{\tilde{X}}(\log D) \oplus \Omega^\bullet_D}{\Omega^\bullet_{\tilde{X}}} \qquad ([3];\ \text{cf. } [1: (7.46)])$$

(here, $\Omega^\bullet_D$ is the direct image of the simplicial sheaf that occurs in the cohomological mixed Hodge complex for $H^\bullet(D)$), or

$$\Omega^\bullet_{\tilde{X}}(\log D) \otimes \mathcal{O}_D \qquad (\text{see } [9: \text{p.}135]).$$

To summarize:

Theorem: Let X be a projective variety with isolated singularities. With notation as above, a Hodge complex for $IH^\bullet(X)$ is

$$C(R\Gamma\Omega^\bullet_{\tilde{X}}(\log D) \longrightarrow \tilde{\tau}^{>m-1}R\Gamma\Omega^\bullet_{\tilde{X}}(\log D) \otimes \mathcal{O}_D).$$

Appendix

A somewhat slicker approach to truncating mixed Hodge complexes is obtained by utilizing the filtration Dec W from [2: (1.3.3)]. Explicitly,

$$(\mathrm{Dec}\ W)_k K^i \;=\; \{x \in W_{k-i}K^i : dx \in W_{k-i-1}K^{i+1}\}.$$

It is, after all, Dec W that canonically induces the weight filtration on the cohomology of a mixed Hodge complex.

We pose the following definition. A <u>shifted mixed Hodge complex</u> (<u>complexe de Hodge mixte décalé</u>) defined over $\mathbb{Q}$ (say) consists of:

i) A filtered complex $(K^\bullet_{\mathbb{Q}}, D_{\mathbb{Q}})$ of vector spaces over $\mathbb{Q}$,

ii) A bifiltered complex $(K^\bullet, D, F)$ of $\mathbb{C}$ vector spaces,

iii) A filtered quasi-isomorphism $(K^\bullet_{\mathbb{Q}} \otimes \mathbb{C}, D_{\mathbb{Q}}) \longrightarrow (K^\bullet, D)$;

such that

iv) $(K^\bullet, D, F)$ is "bistrict" in the sense of Saito (cf. also [8: p.470]), viz. all four spectral sequences

$$\begin{array}{ccccc} & & H(Gr^D Gr^F K^\bullet) & & \\ & \swarrow & & \searrow & \\ H(Gr^F K^\bullet) & & & & H(Gr^D K^\bullet) \\ & \searrow & & \swarrow & \\ & & H(K^\bullet) & & \end{array}$$

degenerate at E_1 (which is a consequence of the degeneration for the two at the right),

v) F induces on

(16) $$Gr^D_k H^i(K^\bullet) \simeq H^i(Gr^D_k K^\bullet)$$

a Hodge structure of weight k for every value of i.

The motivation for the definition is, of course, that given an ordinary mixed Hodge complex $(K^\bullet, W, F)$, putting $D = \mathrm{Dec}\ W$ converts it into a shifted mixed Hodge complex. From Prop. 2, it is evident that the category is closed under truncations.

The concept of purity has an interesting formulation in this new setting. The mixed Hodge structure on $H^i(K^\bullet)$ is pure of weight i if and only if (16) — often denoted also $Gr^W_k H^i(K^\bullet)$, with a built-in shift — vanishes whenever $i \neq k$, i.e. $Gr^D_k K^\bullet$ has all of its cohomology in degree k. When this holds for all k, one can then elect to forget D ....

References

[1] Clemens, C.H.: <u>Degeneration of Kähler manifolds</u>. Duke Math. J. <u>44</u>, 215-290 (1977)

[2] Deligne, P.: Théorie de Hodge, II. Publ. Math. IHES <u>40</u>, 5-57 (1971); III, <u>44</u>, 5-77 (1974)

[3] Durfee, A.: Mixed Hodge structures on punctured neighborhoods. Duke Math. J. <u>50</u>, 1017-1040 (1983)

[4] El Zein, F.: Mixed Hodge structures. Trans. AMS 275, 71-106 (1983)

[5] Hain, R.: The de Rham homotopy theory of complex algebraic varieties, I. To appear.

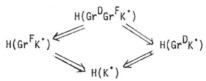

[6] Navarro Aznar, V.: Sur la théorie de Hodge des variétés algébriques à singularités isolées, 1983

[7] Steenbrink, J.: Mixed Hodge structures associated with isolated singularities. In: Singularities, Proc. Symp. Pure Math 40(2), 513-536 (1983)

[8] Zucker, S.: Hodge theory with degenerating coefficients: L_2 cohomology in the Poincaré metric. Ann. Math. 109, 415-476 (1979)

[9] Zucker, S.: Degeneration of Hodge bundles (after Steenbrink). In: Topics in Transcendental Algebraic Geometry. Ann. Math. Studies 106, 121-141 (1984)

Poincaré Lemma for a Variation of Polarized Hodge Structure

Masaki Kashiwara

Research Institute for Mathematical Sciences
Kyoto University
Kyoto, Japan

§1. Introduction

Let X be a compact complex variety, bimeromorphic to a compact Kähler manifold and let $j: X^* \hookrightarrow X$ be the open inclusion from a Zariski open subset X^* of X. The conjecture is that, for any variation of polarized Hodge structure H on X^* (with quasi-unipotent local monodromy), the intermediate cohomology groups $H^n(X; {}^{\pi}H)$ have pure Hodge structure. Here ${}^{\pi}H$ is the minimal extension of H. When $X = X^*$ is a Kähler manifold, this is shown by P. Deligne (See [Z]) and when X is a curve, this is shown by S. Zucker [Z]. Recently, the author, with T. Kawai, proved this when X is a compact Kähler manifold and $X \setminus X^*$ is a normally crossing hypersurface ([K-K]). In this article I report this result. The same result is also obtained by E. Cattani, A. Kaplan and W. Schmid independently ([C-K-S2]. See their article(s) in the same volume). In [K-K2], we give an explicit description of the Hodge filtration of $H^n(X; {}^{\pi}H)$.

§2. Methods

The proof follows basically the proof of Zucker [Z] for the one-dimensional case. However, there are several difficulties caused by a complexity of the behavior of a variation of polarized Hodge structure at the singularities.

The proof proceeds as follows

(2.1) By modifying the Kähler metric on X we take a complete Kähler metric on X^*.

(2.2) We prove that $H^n(X; {}^{\pi}H)$ is isomorphic to the L_2-cohomology groups with respect to the metric introduced in (2.1).

(2.3) By using the harmonic analysis, we represent the L_2-cohomology group as the space of harmonic forms.

(2.4) Decomposing harmonic forms into their (p,q)-components, we obtain the Hodge decomposition of the L_2-cohomology groups.

The most delicate part is (2.2) in which we need the behavior of variation of polarized Hodge structure at singularities.

§3. Variation of polarized Hodge structure

3.1 A Hodge structure H of weight w consists of data $(H_{\mathbb{C}}, F(H), \bar{F}(H))$. Here, $H_{\mathbb{C}}$ is a finite-dimensional $\mathbb{C}$-vector space, $F(H)$ and $\bar{F}(H)$ are finite filtrations of $H_{\mathbb{C}}$ such that $H_{\mathbb{C}} \xleftarrow{\sim} F^p(H) \oplus \bar{F}^q(H)$ for $p+q = w+1$. A polarization S on H is a bilinear form S: $H_{\mathbb{C}} \otimes \overline{H_{\mathbb{C}}} \to \mathbb{C}$ such that

$$(3.1)\quad S(F^p(H), \overline{\bar{F}^q(H)}) = S(\bar{F}^p(H), \overline{F^q(H)}) = 0 \quad \text{for } p+q > w,$$

and

(3.2) $i^{p-q}S(x,\bar{y})$ is a positive definite Hermitian form on $H^{p,q}(H) = F^p(H) \cap \bar{F}^q(H)$ for $p+q = w$.

Here, for a $\mathbb{C}$-vector space V, $\bar{V}$ denotes the complex conjugate of V and $x \mapsto \bar{x}$ is the $\mathbb{R}$-linear isomorphism from V to $\bar{V}$ such that $\overline{ax} = \bar{a}\bar{x}$ for $a \in \mathbb{C}$, $x \in \bar{V}$.

3.2 Let X be a complex manifold, and $\bar{X}$ the complex conjugate of X. Therefore, the structure sheaf $\underline{O}_{\bar{X}}$ of $\bar{X}$ is the sheaf of anti-holomorphic functions on X. A variation of Hodge structure H of weight w consists of data $(H_{\mathbb{C}}, F(H), \bar{F}(H))$, where $H_{\mathbb{C}}$ is a locally constant sheaf of $\mathbb{C}$-vector spaces of finite dimension on X and $F(H)$ (resp. $\bar{F}(H)$) is a filtration of $\underline{O}_X \underset{\mathbb{C}}{\otimes} H_{\mathbb{C}}$ (resp. $\underline{O}_{\bar{X}} \underset{\mathbb{C}}{\otimes} H_{\mathbb{C}}$) by vector subbundles such that, at any $x \in X$, $H(x) = (H_{\mathbb{C},x}, F(H)(x), \bar{F}(H)(x))$ is a Hodge structure of weight w and that $vF^p(H) \subset F^{p-1}(H)$ (resp. $v\bar{F}^p(H) \subset \bar{F}^{p-1}(H)$) for any p and any holomorphic vector field v on X (resp. $\bar{X}$). A polarization S of H is a homomorphism $S\colon H_{\mathbb{C}} \otimes \overline{H_{\mathbb{C}}} \longrightarrow \mathbb{C}_X$ such that S_x is a polarization of $H(x)$ for any $x \in X$.

§4. Nilpotent orbit theorem

4.1 Let H be a finite-dimensional complex vector space, w an integer and let $S\colon H \otimes \bar{H} \to \mathbb{C}$ be a non-degenerate bilinear form such that $\overline{S(x,y)} = (-1)^w S(\bar{y},\bar{x})$ for any $x \in H$ and $y \in \bar{H}$. Let $\underline{u}(S) = \{A \in \underline{gl}(H);\ S(Ax,\bar{y})+S(x,\overline{Ay}) = 0$ for any $x,y \in H\}$.

Let F and $\bar{F}$ be filtrations of H satisfying (3.1). Let $\{N_1,\cdots,N_n\}$ be a mutually commuting set of nilpotent elements of $\underline{u}(S)$ such that $N_jF^p \subset F^{p-1}$ and $N_j\bar{F}^p \subset \bar{F}^{p-1}$. Then the following conditions (i) and (ii) are equivalent (the main part is due to Cattani-Kaplan [C-K]).

(i) The N-filtration $W(N)$ does not depend on $N \in C(N_1,\ldots,N_n) = \{\sum_{j=1}^{n} t_jN_j;\ t_j > 0\}$ and F, $\bar{F}$ induce a pure Hodge structure of weight $w+k$ on $Gr_k^{W(N)}$ for any k. Moreover, $S(x, \overline{N^k y})$ gives a polarization on $P_k(N) = \mathrm{Ker}(N^{k+1}\colon Gr_k^{W(N)} \to Gr_{-k-2}^{W(N)})$ for $k \geq 0$.

(ii) There exists $N_0 \in C(N_1,\cdots,N_n)$ such that, for any $N \in N_0 + C(N_1,\cdots,N_n)$, $(e^{iN}F, e^{-iN}\bar{F})$ is a Hodge structure of weight w and S polarizes it.

If these equivalent two conditions are satisfied, we say that $(H; S; F, \bar{F}; N_1,\cdots,N_n)$ forms a nilpotent orbit of weight w. We denote by $W(N_1,\cdots,N_n)$ the N-filtration for $N \in C(N_1,\cdots,N_n)$.

4.2 Now let $(H; S; F, \bar{F}; N_1, \cdots, N_n)$ form a nilpotent orbit. Let $H_{\mathbb{C}}$ be the local system on $X^* = \mathbb{C}^{*n}$ with H as a stalk and with $\exp N_j$ as the local monodromy around $z_j = 0$. Therefore defining the integrable connection

$$d: \underline{O}_X \otimes H \longrightarrow \Omega^1_{X^*} \otimes H$$

by $de = -\sum_j N_j e \dfrac{dz_j}{2\pi\sqrt{-1}z_j}$, $H_{\mathbb{C}}$ is isomorphic to the sheaf of flat section of H. Then $\underline{O}_{X^*} \otimes H \cong \underline{O}_{X^*} \otimes H_{\mathbb{C}}$ and $F^p(H)$ give the filtration on $\underline{O}_{X^*} \otimes H_{\mathbb{C}}$. Similarly $\bar{F}(H)$ gives the filtration of $\underline{O}_{\bar{X}^*} \otimes H_{\mathbb{C}}$. Then $H_{\mathbb{C}}$ is a variation of polarized Hodge structure on a neighborhood of the origin. Conversely, any variation of polarized Hodge structure on $U \cap X^*$ with a neighborhood U of the origin is approximated by the variation coming from nilpotent orbits. This is what Schmid's nilpotent orbit theorem ([S]) says. Therefore, the study of behavior of variation of polarized Hodge structure with singularities often reduces to the study of nilpotent orbits.

4.3 The Vanishing Cycle Theorem. Let $(H; S; F, \bar{F}; N_1, \cdots, N_n)$ form a nilpotent orbit of weight w. Set $H_1 = \operatorname{Im} N_1$ and let us define $S_1: H_1 \otimes H_1 \longrightarrow \mathbb{C}$ by $S_1(N_1x, \overline{N_1y}) = S(x, \overline{N_1y})$. Then $(H_1; S_1; N_1F, N_1\bar{F}; N_1|H_1, \cdots, N_n|H_1)$ forms a nilpotent orbit of weight $w+1$ and $W_k(N_1|H_1, \cdots, N_n|H_1) = H_1 \cap W_{k-1}(N_1, \cdots, N_n) = N_1 W_{k+1}(N_1, \cdots, N_n)$.

The meaning of this theorem is as follows. Let H be a variation of polarized Hodge structure on Δ^{*n} where $\Delta^* = \{z \in \mathbb{C};\ 0<|z|<1\}$. Then the vanishing cycle sheaf of ${}^{\pi}H$ with respect to x_1 is also a variation of polarized Hodge structure.

4.4 Let $(F, \bar{F}; N_1, \cdots, N_n)$ form a nilpotent orbit. Letting $\{e_1, \cdots, e_n\}$ be a base of $\mathbb{Z}^n$, we define the partial Koszul complex $\Pi(N_1, \cdots, N_n)$ by

$$\Pi(N_1, \cdots, N_n)^k = \bigoplus_{|J|=k} (\operatorname{Im} N_J) e_J \subset H \otimes \bigwedge^k \mathbb{Z}^n.$$

Here J ranges over the set of subsets of $\{1, \cdots, n\}$ with k elements and $N_J = \prod_{j \in J} N_j$ and $e_J = \bigwedge_{j \in J} e_j$. The differential is given by the exterior multiplication of $\sum_{j=1}^{n} N_j e_j$. We endow $\Pi(N_1, \cdots, N_n)$ with the mixed Hodge structure by

$$W_k(\mathrm{Im}\, N_J) = N_J W_{k-w}(N_1,\cdots,N_n)$$

$$F^p(\mathrm{Im}\, N_J) = N_J F^p$$

$$\bar{F}^p(\mathrm{Im}\, N_J) = N_J \bar{F}^p.$$

Then $\Pi(N_1,\cdots,N_n)$ is a complex in the abelian category of mixed Hodge structures.

The Purity Theorem. If $(F, \bar{F}; N_1,\cdots,N_n)$ forms a nilpotent orbit of weight w, then $H^k(\Pi(N_1,\cdots,N_n))$ has weight $\leqq w+k$.

Remark that

$$H^0(\Pi(N_1,\cdots,N_n)) = \bigcap_{j=1}^{n} \mathrm{Ker}\, N_j$$

$$H^k(\Pi(N_1,\cdots,N_n)) = 0 \quad \text{for} \quad k \geqq n$$

$$H^{n-1}(\Pi(N_1,\cdots,N_n)) = \mathrm{Ker}\, N_1\cdots N_n \Big/ \sum_{j=1}^{n} \mathrm{Ker}(\prod_{k\neq j} N_k).$$

So the purity theorem implies

$$\text{(4.4.1)} \qquad \mathrm{Ker}\, N_1\cdots N_n \subset \sum_{j=1}^{n} \mathrm{Ker} \prod_{k\neq j} N_k + W_{n-1}(N_1,\cdots,N_n)$$

By the duality, we have

$$\text{(4.4.2)} \qquad \bigcap_{j=1}^{n} \mathrm{Im} \prod_{k\neq j} N_k \cap W_{-n}(N_1,\cdots,N_n) \subset \mathrm{Im}\, N_1\cdots N_n.$$

4.5 Let H be a variation of polarized Hodge structure of weight w on $X^* = \Delta^{*n}$, and let ${}^{\pi}H_{\mathbb{C}}$ be the minimal extension of $H_{\mathbb{C}}$ to $X = \Delta^n$. Here Δ is the unit disc and $\Delta^* = \Delta\setminus\{0\}$.

Letting N_j be the logarithm of the unipotent monodromy around $z_j = 0$, $\Pi(N_1,\cdots,N_n)$ is quasi-isomorphic to the germ ${}^{\pi}H_0$ of ${}^{\pi}H$ at the origin. Thus, the purity theorem says that $H^k({}^{\pi}H_0)$ has weight $\leqq w+k$. On the other hand, $H^k(X\setminus\{0\}; H)$ is isomorphic to $H^k({}^{\pi}H_0)$ for $k < n$, and $(H^{2n-k-1}({}^{\pi}(H^*(n))_0))^*$ for $k \geqq n$. This shows

(4.5.1) The weight λ of $H^k(X\setminus\{0\}; H)$ satisfies $\lambda \leqq w+k$ (for $k < n$) and $\lambda \leqq w+k+1$ (for $k \geqq n$).

§5. L_2-Cohomology

5.1 Let X be a complex manifold and X^* an open subset of X such that $X \backslash X^*$ is a normally crossing hypersurface. Let H be a variation of polarized Hodge structure of weight w on X^* with quasi-unipotent local monodromies around $X \backslash X^*$.

5.2 Let us take a Riemannian metric g on X^* such that for any $p \in X \backslash X^*$ and a local coordinate system $(z_1, \cdots, z_n)$ around p with $X^* = \{z;\ z_1 \cdots z_\ell \neq 0\}$, we have

$$(5.2.1) \qquad g \sim \sum_{j \leq \ell} \frac{|dz_j|^2}{(|z_j| \log |z_j|)^2} + \sum_{j > \ell} |dz_j|^2 .$$

Here $\sim$ means that each side is majorated by a constant times the other.

5.3 Let us denote by $\underline{E}$ the sheaf of C^∞ functions on X^*, $\underline{E}^{(k)}$ the sheaf of C^∞ k-forms and $\underline{E}^{(p,q)}$ the sheaf of C^∞ (p,q)-forms on X^*. Let $\underline{Db}$ denote the sheaf of distributions on X^*. We set

$$F^p(\underline{E}^{(k)} \otimes H_{\mathbb{C}}) = \bigoplus_{k=p'+q'} \underline{E}^{(p',q')} \underset{\underline{O}_{X^*}}{\otimes} F^{p-p'}(H)$$

and

$$\bar{F}^q(\underline{E}^{(k)} \otimes H_{\mathbb{C}}) = \bigoplus_{k=p'+q'} \underline{E}^{(p',q')} \underset{\underline{O}_{\bar{X}^*}}{\otimes} \bar{F}^{q-q'}(H).$$

Let $\underline{L}_2^k(H)$ denote the sheaf on X defined as follows. For any U, $\Gamma(U; \underline{L}^k(H))$ consists of sections u of $\underline{Db} \underset{\underline{E}}{\otimes} \underline{E}^{(k)} \otimes H$ over $U \cap X^*$ such that $u|_{V \cap X^*}$ and $du|_{V \cap X^*}$ are square integrable for any relatively compact open subset V of U. Here the fiber metric of H is given by the polarization and the fiber metric of $\underline{E}^{(p)}$ and the volume elements of X^* come from the Riemannian metric g on X^*. Then $\underline{L}_2^{\cdot}(H)$ forms a complex of sheaves on X. Furthermore, letting j be the inclusion $X^* \hookrightarrow X$, $F^p(\underline{L}_2^k(H)) = \underline{L}_2^k(H) \cap j_*(\underline{Db} \otimes F^p(\underline{E}^{(k)} \otimes H))$ defines the filtration of the complex $\underline{L}_2^{\cdot}(H)$. Similarly, we can introduce the filtration $\bar{F}(\underline{L}_2^{\cdot}(H))$ of $\underline{L}_2^{\cdot}(H)$.

Theorem. (i) $\underline{L}_2^{\cdot}(H)$ *is a soft sheaf.*

(ii) $\underline{L}_2^\cdot(H)$ *is the minimal extension of* H; i.e. $\underline{L}_2^\cdot(H)$ *is a perverse complex* (See e.g. [BBD]) *on* X^* *such that*

(5.3.1) $\underline{L}_2^\cdot(H)|_{X^*} \cong H_{\mathbb{C}}$ *(quasi-isomorphic) and*

(5.3.2) $\underline{L}_2^\cdot(H)$ *has neither non-zero subobjects nor non-zero quotients supported in* $X\setminus X^*$ *in the category of perverse complexes.*

The proof uses the asymptotic behavior of the polarization at the singularities ([K], [C-K-S]) and the purity theorem, as explained in the next section.

§6. Outline of the proof of Theorem

6.1 We shall give a sketch of the proof of Theorem in §5.

The question being local, we may assume $X = D^n$, $X^* = D^{*n}$ where D is the unit disc and $D^* = D\setminus\{0\}$. By the induction on n, we may assume

(6.1.1) $\underline{L}_2^\cdot(H)|_{X\setminus\{0\}} \cong {}^{\pi}H_{\mathbb{C}}|_{X\setminus\{0\}}$.

Thus, by the characterization of ${}^{\pi}H_{\mathbb{C}}$, it is enough to show

(6.1.2) $$H^j(X;\underline{L}_2^\cdot(H)) = \begin{cases} H^j(X\setminus\{0\};\underline{L}_2^\cdot(H)) & \text{for } j < n \\ 0 & \text{for } j \geqq n. \end{cases}$$

6.2 We shall use the coordinates $(t,y_1,\cdots,y_n,x_1,\cdots,x_n)$ of X^* where $(z_1,\cdots,z_n) \in D^{*n} = X$, $z_j = \exp(2\pi\sqrt{-1}\, x_j - \frac{1}{t}y_j)$ with $\sum y_j = 1$, $y_j, t > 0$ and $x_j \in \mathbb{R}/\mathbb{Z}$.

Let $X^*(t)$ be the fiber of $X^* \ni (x,y,t) \longmapsto t \in \mathbb{R}^+$.

Then we have

(6.2.1) $$H^j(X;\underline{L}_2^\cdot(H)) = H^j(L_2^\cdot(X^*;H))$$

and

(6.2.2) $$H^j(X\setminus\{0\};\ \underline{L}_2^\cdot(H)) = H^j(L_2^\cdot(X^*(t_0);H)) \quad \text{for some } 0 < t_0 \ll 1.$$

Here, $L_2^{\cdot}(X^*;H)$ denotes the space of H-valued forms such that u and du are square integrable.

Hence we reduced the problem to show

$$(6.2.3)\qquad H^j(L_2^{\cdot}(X^*;H)) = \begin{cases} H^j(L^2(X^*(t_0);H)) & \text{for } j < n \\ 0 & \text{for } j \geqq n. \end{cases}$$

<u>6.3</u> We can consider an element in $L_2^{\cdot}(X^*;H)$ as an $L_2^{\cdot}(X^*(t_0);H)$-valued forms in t, with certain growth condition. To describe this we need the following result in [K] and [C-K-S] concerning the asymptotic behavior of polarization near the boundary.

<u>6.4</u> Let N_j be the logarithm of the unipotent part of the monodromy around $z_j = 0$.

Let W be the monodromy weight filtration of $H_{\mathbb{C}}$ with respect to $\sum N_j$. We decomposes $\underline{O}_{X^*}\otimes H_{\mathbb{C}}$ as $\oplus I_k$ such that $Gr_k^W \cong I_k$. Let Y be the weight operator given by $Y|_{I_k} = k$. Let $\| \quad \|_{(t,x,y)}$ be the metric of $H_{\mathbb{C}}$ induced from the polarization at the point (t,x,y) of X*. Then we have

<u>Proposition</u> 6.4.1. ([K] [C-K-W]) <u>For</u> $u \in H_{\mathbb{C}}$, <u>we have</u>

$$\|u\|_{(t,x,y)} \sim \|t^{-Y/2}u\|_{(t_0,x,y)}.$$

<u>6.5</u> Now, we have

$$(6.5.1)\qquad H^j(L_2^{\cdot}(X^*(t_0);H)) = H^j(X\setminus\{0\};\ {}^{\pi}H)$$

and hence it is finite-dimensional. Hence if $\underline{h}_k$ denotes the space of harmonic k-forms belonging to $L_2^{\cdot}(X^*(t_0);H)$, then

$$(6.5.2)\qquad \underline{h}_k \cong H^k(L_2^{\cdot}(X^*(t_0);H)).$$

Note that the weight operator Y operates also on $\underline{h}_k$.

<u>6.6</u> Let $S^{\cdot}$ be the subcomplex of $L^2(X^*;H)$ consisting of $(\underset{k}{\oplus}\underline{h}_k)$-valued forms in t.

Then we can show by using (6.5.2)

(6.6.1) $H^j(L^2(X^*;H)) = H^j(S^\cdot)$.

Now for an $\underline{h}_k$-valued function $u(t)$, Proposition 6.4.1 gives

(6.6.2) $\|u\|^2 \sim \int \|t^{-Y/2}u(t)\|^2_{\underline{h}_k} \frac{dt}{t}$.

(6.6.3) $\|\frac{dt}{t}u\|^2 \sim \int \|t^{-Y/2}u(t)\|^2_{\underline{h}_k} \frac{dt}{t}$.

Then Theorem follows from the following Hardy's inequality (6.6.4) and (6.6.5) which follows from (4.5.1).

(6.6.4) Let $r \in \mathbb{R}\setminus\{0\}$. Then for any $f(t)$, there is $F(t)$ such that $t\frac{dF(t)}{dt} = f(t)$ and $\int_0^\infty t^r |F(t)|^2 \frac{dt}{t} \leqq (\frac{2}{r})^2 \int_0^\infty t^r |f(t)|^2 dt/t$.

(6.6.5) Any eigen-value λ of $Y|_{\underline{h}_k}$ satisfies $\lambda \leqq k$ (when $k < n$) $\lambda \geqq k+1$ (when $k \geqq n$).

In fact, they imply

(6.6.6) $$H^j(S^\cdot) \cong \begin{cases} \underline{h}_j & \text{for } j < n \\ 0 & \text{for } j \geqq n. \end{cases}$$

Then (6.6.6), (6.6.1) and (6.5.2) give the desired result (6.2.3).

§7. Hard Lefschetz theorem

7.1 Now assume that (X,ω) is a compact Kähler manifold. Then we can choose $\omega + \sqrt{-1}\,\partial\bar{\partial}\log|\log\varphi|$ as g for a C^∞-function φ vanishing on $X\setminus X^*$. Then g is also a Kähler metric.

Since $H^n(\Gamma(X;\underline{L}_2^\cdot(H))) \cong H^n(X;{}^\pi H)$ is finite-dimensional, we can use the harmonic analysis and we can argue similarly to Weil's book [W]. The cohomology group $H^n(\Gamma(X;\underline{L}_2^\cdot(H)))$ is isomorphic to the space of harmonic n-forms, and their decomposition to the (p,q)-components gives a Hodge decomposition of the cohomology group. Summing up we obtain the following theorem.

7.2 Theorem. Let X be a complex Kähler manifold of dimension n, $X \setminus X^*$ a normally crossing hypersurface and H a variation of polarized Hodge structure of weight w on X^*. Then we have

(i) $H^k(X;F^p(\underline{L}_2^{\bullet}(H))) \to H^k(X;\underline{L}_2^{\bullet}(H))$ and $H^k(X;\bar{F}^p(\underline{L}_2^{\bullet}(H))) \to H^k(X;\underline{L}_2^{\bullet}(H))$ are injective and their images give the pure Hodge structure of weight $w+k$ on $H^k(X;\underline{L}_2^{\bullet}(H)) \cong H^k(X;{}^{\pi}H_{\mathbb{C}})$.

(ii) Let $\ell = [\omega] \in H^2(X;\mathbb{R})$ be the cohomology class of the Kähler form ω. Then

$$\ell^k: H^{n-k}(X;{}^{\pi}H_{\mathbb{C}}) \longrightarrow H^{n+k}(X;{}^{\pi}H_{\mathbb{C}})$$

is an isomorphism for $k \geqq 0$.

(iii) Letting P_k be the kernel of $\ell^{k+1}: H^{n-k}(X;{}^{\pi}H_{\mathbb{C}}) \to H^{n+k+2}(X;{}^{\pi}H_{\mathbb{C}})$, $(-1)^{(n-k)(n-k-1)/2}(\alpha,\ell^k\beta)$ gives a polarization of P_k. Here, $(*,*)$ is given by $H^{n-k}(X;{}^{\pi}H_{\mathbb{C}}) \otimes H^k(X;{}^{\pi}H_{\mathbb{C}}) \to H^{2n}(X;{}^{\pi}H_{\mathbb{C}} \otimes {}^{\pi}H_{\mathbb{C}}) \to H^{2n}(X;\mathbb{C}) \to \mathbb{C}$.

References

[B-B-D] A. A. Beilinson, J. Bernstein and P. Deligne, Faisceaux pervers, Astérisque, 100 (1982), Soc. Math. France.

[C-K] E. Cattani and A. Kaplan, Polarized mixed Hodge structures and the local monodromy of a variation of Hodge structures, Inv. Math. 67(1982), 101-115.

[C-K-S] E. Cattani, A. Kaplan and W. Schmid, The SL_2-orbit theorem in several variables (to appear).

[C-K-S2] E. Cattani, A. Kaplan and W. Schmid, L_2 and intersection cohomologies for a polarizable variation of Hodge structure, preprint.

[K] M. Kashiwara, The asymptotic behavior of a variation of polarized Hodge structures, Publ. of R.I.M.S., Kyoto Univ. 21(1985), 853-875.

[K-K] M. Kashiwara and T. Kawai, The Poincaré lemma for a variation of polarized Hodge structure, Proc. Japan Acad., 61 Ser. A(1985), 164-167.

[K-K2] M. Kashiwara and T. Kawai, Hodge structure and holonomic systems, Proc. Japan Acad. 62, Ser. A (1986) 1-4.

[K-K3] M. Kashiwara and T. Kawai, The Poincaré lemma for variations of Hodge structure, to appear in Publ. R.I.M.S.

[S] W. Schmid, Variation of Hodge structure: the singularities of the period mappings, Inv. Math. 22 (1973), 211-319.

[W] A. Weil, Introduction à l'étude des variétés kähleriennes, Hermann, Paris, 1958.

[Z] S. Zucker, Hodge theory with degenerating coefficients, Annals of Math., 109 (1979), 415-476.

EVALUATION D'INTEGRALES ET THEORIE DE HODGE

F. LOESER

Centre de Mathématiques de l'Ecole Polytechnique
Plateau de Palaiseau - 91128 Palaiseau - Cedex
"U.A. du C.N.R.S. n° 169"

I - INTRODUCTION

Soit p un nombre premier, $\mathbb{Z}_p$ l'anneau des entiers p-adiques, $\mathbb{Q}_p$ le corps des nombres p-adiques.

Soit $f_1,\dots,f_r$ des polynômes en m variables à coefficients dans $\mathbb{Z}_p$.

Pour tout entier n notons :

$\widetilde{N}_n$ le cardinal de $\{x \bmod p^n \mid x \in \mathbb{Z}_p^m$ et $f_i(x) \equiv 0 \bmod p^n$ pour $1 \leqslant i \leqslant r\}$ (solutions approchées) et N_n le cardinal de $\{x \bmod p^n \mid x \in \mathbb{Z}_p^m$ et $f_i(x) = 0$ pour $1 \leqslant i \leqslant r\}$ (solutions exactes).

On considère les séries de Poincaré

$$\widetilde{P}(T) = \sum_{n=o}^{\infty} \widetilde{N}_n T^n \quad \text{et} \quad P(T) = \sum_{n=o}^{\infty} N_n T^n .$$

Borevitch et Shafarevitch ont conjecturé que $\widetilde{P}(T)$ est une fonction rationnelle de T. Ceci a été montré par J. Igusa [I1] pour $r = 1$, en utilisant le théorème de résolution des singularités d'Hironaka et par Meuser [M1] pour r quelconque.

La question de la rationalité de $P(T)$ a été posée par Serre [Se] et résolue par J. Denef dans [D1].

Pour montrer la rationalité de ces séries de Poincaré on les exprime comme intégrales grâce aux relations suivantes :

Relation I.1 (Igusa [I1], Meuser [M1]) :

$$\widetilde{P}(p^{-m-s}) = \frac{1-p^{-s}\ \widetilde{I}(s)}{1-p^{-s}}$$

avec $$\widetilde{I}(s) = \int_{\mathbb{Z}_p^m} |f(x)|_p^{s}\ |dx|_p$$

et $|f(x)|_p = \underset{i}{\mathrm{Max}} |f_i(x)|_p$ pour $\mathrm{Re}\ s > 0$.

($| \ |_p$ est la norme standard sur $\mathbb{Q}_p$ et $|dx|_p$ la mesure de Haar standard sur $\mathbb{Q}_p^m$).

Relation I.2 (Denef [D1]) :

$$P(p^{-m-s}) = \frac{1-p^{-s}I(s)}{1-p^{-s}}$$

avec $I(s) = \int_{\mathbb{Z}_p^m} d(x,V)^s \, |dx|_p$, pour $\mathrm{Re}\ s > 0$,

V désignant l'ensemble analytique défini par

$$f_i(x) = 0 \quad 1 \leqslant i \leqslant r \quad \text{et} \quad d(x,V) = \inf(d(x,y)/y \in \mathbb{Z}_p^m \cap V)$$

Il reste alors à montrer que $\tilde{I}(s)$ et $I(s)$ se prolongent analytiquement à $\mathbb{C}$ en une fonction rationnelle de p^{-s} . C'est ce que fait J. Igusa pour $\tilde{I}(s)$ dans [I1] dans le cas $r = 1$ en utilisant le théorème d'Hironaka pour rendre l'hypersurface $f_1 = 0$ à croisements normaux. Pour traiter le cas de $I(s)$, J. Denef doit utiliser le théorème d'élimination des quantificateurs de Macintyre, analogue p-adique du théorème de Tarski-Seidenberg car la fonction $d(x,V)$ est seulement semi-algébrique (au sens de [D4]). En utilisant ce résultat il obtient une preuve de la rationalité de $\tilde{P}$ et P n'utilisant pas la résolution des singularités ([D1]).

Nous nous proposons d'exposer dans le présent texte les résultats que nous avons obtenus dans l'étude des pôles des analogues complexes des intégrales $\tilde{I}(s)$ et $I(s)$ en utilisant des méthodes transcendantes comme la théorie de Hodge, et comment certains de ces résultats peuvent avoir des conséquences dans le cas non archimédien.

Pour d'autres aspects arithméthiques de la théorie des "puissances complexes" d'Igusa, concernant en particulier la formule de Siegel-Weil et ses généralisations nous renvoyons aux articles d'Igusa [I1-11], et plus particulièrement aux beaux textes d'introduction [I5] et [I9]. Pour l'utilisation de cette théorie dans un contexte adélique nous renvoyons à [Da].

II - POLES DE $|f|^{2s}$

II.1 Pôles de $|f|^{2s}$ et racines du polynôme de Bernstein-Sato

Soit $f : (\mathbb{C}^{n+1},0) \to (\mathbb{C},0)$ un germe de fonction analytique. On note $\tilde{f}$ un représentant de f , $X = \tilde{f}^{-1}(D_\eta) \cap B_\varepsilon$ un tube de Milnor ($0 < \eta \ll \varepsilon \ll 1$; B_ε est la boule de centre zéro et rayon ε , D_η le disque de centre 0 et rayon η).

On note $f : X \to \Delta$ la restriction de $\widetilde{f}$ à X, avec $\Delta = D_\eta$.

Soit ψ une $(n+1\ ,\ n+1)$ forme sur $\mathbb{C}^{n+1}$ à support contenu dans X. Pour $s \in \mathbb{C}$ avec $\operatorname{Re} s > 0$ on considère l'intégrale :

$$I(\psi,s) = \int_X |f|^{2s}\ \psi$$

L'existence du polynôme de Bernstein-Sato dont nous allons rappeler la définition permet de montrer facilement l'existence du prolongement analytique de $I(\psi,s)$ en une fonction méromorphe sur $\mathbb{C}$ et d'avoir une première estimation sur l'ensemble de ses pôles.

Théorème et Définition II.1.1 ([Be]) .

1) Il existe un opérateur différentiel analytique à coefficients polynomiaux en s, $P(s)$, et un polynôme non nul $b \in \mathbb{C}[s]$ tels que :

$$(*) \qquad P\,f^s = b\,f^{s-1}$$

(ici f^s n'est pas vue comme une fonction, mais comme un générateur formel d'un certain module sur les opérateurs différentiels à coefficients polynomiaux en s (cf. [Be])).

2) Le générateur monique de l'idéal des $b \in \mathbb{C}[s]$ pour lesquels il existe P vérifiant (*) est appelé le polynôme de Bernstein-Sato de f et est noté b. Comme $b(0) = 0$ (faire $s=0$ dans (*)), on peut poser $b = s\,\widetilde{b}$ avec $\widetilde{b} \in \mathbb{C}[s]$.

En utilisant l'équation (*) on obtient facilement le résultat suivant :

Théorème II.1.2

1) $I(\psi,s)$ se prolonge analytiquement en une fonction méromorphe sur $\mathbb{C}$.

2) Soit $P(|f|^{2s},\psi)$ l'ensemble des pôles de $I(\psi,s)$, on a :

$$P(|f|^{2s},\psi) \subset \{s-m / b(s) = 0 \text{ et } m \in \mathbb{N}^*\}$$

L'application $\psi \to I(\psi,s)$ définit un courant noté $|f|^{2s}$ dépendant méromorphiquement du paramètre s. On note $P(|f|^{2s})$ l'ensemble de ses pôles. On a par définition : $P(|f|^{2s}) = \bigcup_\psi P(|f|^{2s},\psi)$. On dira que α est un pôle de $|f|^{2s}$ d'ordre au moins k si il existe ψ telle que α est pôle de $I(\psi,s)$ d'ordre au moins k.

Nous nous intéressons à la question suivante, réciproque de II.1.2 : si s est une racine du polynôme de Bernstein-Sato de f, $s-1$ est-il nécessairement pôle de

$|f|^{2s}$? Comme les entiers strictement négatifs sont toujours des pôles (triviaux) de $|f|^{2s}$, même dans le cas où f n'a pas de singularités (ils sont alors simples), on peut se demander si de plus, quand s est une racine entière du polynôme de Bernstein-Sato réduit $\tilde{b}$, alors nécessairement $s-1$ est pôle d'ordre au moins deux de $|f|^{2s}$.

Nous avons montré que la réponse à ces questions est affirmative dans le cas où f définit une courbe réduite ou une singularité quasi-homogène isolée, et donc dans ce cas on a entièrement déterminé $P(|f|^{2s})$:

Théorème II.1.3 ([L2])

Si f est à singularité isolée en zéro et définit, soit une courbe $(n=1)$ soit une singularité quasi-homogène, on a $P(|f|^{2s}) = \{s-m \mid b(s) = 0$ *et* $m \in \mathbb{N}^*\}$.

De plus si s est une racine entière de $\tilde{b}$ le pôle en $s-1$ est d'ordre au moins deux.

Pour une singularité isolée générale on a le résultat plus partiel suivant :

Théorème II.1.4 ([L2])

Si f est à singularité isolée en zéro, alors

$$E = \{-(\alpha+m) \;/\; \alpha \in Sp(f)\;,\; m \in \mathbb{N}^*\} \subset P(|f|^{2s})\;,$$

avec $Sp(f)$ le spectre de Steenbrink-Varchenko (défini dans [St2], [V2]). De plus tout point entier de E est un pôle d'ordre au moins deux.

Comme d'après Malgrange ([Ma]), $\sup\{\alpha \mid -\alpha \in Sp(f)\} = \sup\{s \mid \tilde{b}(s) = 0\}$ on a le corollaire suivant de II.1.4 :

Corollaire II.1.5 : *Si β est la plus grande racine de $\tilde{b}$, $\beta-1$ est le plus grand pôle non trivial de $|f|^{2s}$. (Les pôles triviaux de $|f|^{2s}$ sont les pôles simples entiers).*

Remarque II.1.6 : Comme dans une déformation à μ constant $B = \{s-m \mid b(s) = 0$ et $m \in \mathbb{N}^*\}$ peut varier d'après les exemples de Yano et Kato, on en déduit d'après II.1.3 que $P(|f|^{2s})$ peut également varier à μ constant (même pour des courbes à une paire).

Remarque II.1.7 : Dans le cas des singularités *non* isolées Barlet a obtenu des résultats analogues aux nôtres mais moins précis. Voici un exemple du type de résultats qu'il obtient : "si la monodromie agissant sur la cohomologie de degré $p \geqslant 1$ de la fibre de Milnor admet la valeur propre $e^{-2\pi i u}$ avec $0 \leqslant u < 1$ alors $-p-u \in P(|f|^{2s})$". Pour plus de détail voir [B1], [B2], [B3], [B4].

D'autre part J. Igusa a obtenu indépendamment des résultats (moins précis que les nôtres) dans le cas des singularités isolées dont la monodromie n'admet pas la valeur propre 1 ([I7], [I8]). Barlet et Igusa ont aussi des résultats sur le lien entre la taille des blocs de Jordan et la multiplicité des pôles. Dans le cas des singularités isolées notre méthode permet également de retrouver ces résultats.

Nous allons maintenant indiquer l'idée de la preuve des théorèmes II.1.3 et II.1.4. Pour cela nous allons rappeler la définition de la forme hermitienne canonique de Barlet ([B5]) :

On note $X(t) = X \cap f^{-1}(t)$ et $\Delta^* = \Delta \setminus \{0\}$.

Sur Δ^* on dispose du fibré de Milnor $\underline{H}$ dont la fibre en t est $H^n(X(t), \mathbb{C})$.

Soit $\pi : U \to \Delta^*$ le revêtement universel de Δ^*, on note $H = \Gamma(U, \pi^* \underline{H})$ et H^* le dual de H.

Si Nilss désigne l'espace vectoriel des fonctions de classe de Nilsson sur U, et $E \subset \text{Nilss} \otimes_{\mathbb{C}} H$ le sous-espace des invariants par la monodromie, on a un morphisme canonique :

$$s : \Gamma(B_\varepsilon, \Omega^{n+1}) \to E,$$

défini ainsi : pour $\gamma \in H^*$ on a

$$s(\omega)[\gamma] = \int_\gamma \frac{\omega}{df}$$

(Ω^{n+1} est le faisceau des $n+1$ formes holomorphes sur $\mathbb{C}^{n+1}$).

Si $\frac{2\pi}{i} N$ désigne le logarithme de la partie unipotente de la monodromie, on peut écrire

$$s(\omega) = \sum_{\alpha \in F+\mathbb{N}} t^\alpha t^N u_\alpha^\omega \quad \text{avec } u_\alpha^\omega \in H$$

où F est un ensemble fini de rationnels, et on note

$$\alpha(\omega) = \inf\{\alpha \mid u_\alpha^\omega \neq 0\}.$$

Soit $\mathcal{N}$ le $\mathbb{C}[[t,\bar{t}]]$ module

$$\bigoplus_{\substack{r \in \mathbb{Q} \\ j \in [0,n]}} \mathbb{C}[[t,\bar{t}]] \, |t|^r (\log t\bar{t})^j / \mathbb{C}[[t,\bar{t}]]$$

Dans [B1] D. Barlet a montré que l'on a un morphisme canonique :

$$\mathcal{H} : \Gamma(B_\varepsilon, \Omega^{n+1}) \times \Gamma(B_\varepsilon, \Omega^{n+1}) \to \mathcal{N}$$

qui à deux $(n+1)$ formes ω et ω' associe le développement asymptotique de

$$\frac{1}{(2\pi i)^n} \int_{X(t)} \frac{\omega}{df} \wedge \frac{\overline{\omega'}}{df} \quad \text{dans } \mathcal{N} .$$

Dans [B5] D. Barlet montre qu'il existe une forme hermitienne h sur H appelée forme hermitienne canonique telle que :

$$\mathcal{H}(\omega,\omega') = \sum_{\substack{\alpha \notin \mathbb{N} \\ \alpha' \notin \mathbb{N} \\ k\in[0,N]}} t^{\alpha}\, \bar{t}^{\alpha'} \frac{(\log t\bar{t})^k}{k!}\, h\,(N^k u_{\alpha}^{\omega}, u_{\alpha'}^{\omega'})$$

$$+ \sum_{\substack{\alpha \in \mathbb{N} \\ \alpha' \in \mathbb{N} \\ k\in[0,N]}} t^{\alpha}\, \bar{t}^{\alpha'} \frac{(\log t\bar{t})^{k+1}}{(k+1)!}\, h(N^k u_{\alpha}^{\omega}, u_{\alpha'}^{\alpha'}) .$$

si $s(\omega) = \Sigma\, t^{\alpha} t^{N} u_{\alpha}^{\omega}$

et $s(\omega') = \Sigma\, t^{\alpha'} t^{N} u_{\alpha'}^{\omega'}$.

Pour λ une valeur propre de la monodromie on note H_{λ} le sous-espace propre généralisé qui lui est associé, et $H_{\neq 1} = \bigoplus_{\lambda \neq 1} H_{\lambda}$.

Dans [B5], D. Barlet montre le résultat suivant :

<u>Théorème II.1.8</u> : 1) <u>h est non dégénérée</u>

2) <u>si Q désigne la forme d'intersection sur $H_{\neq 1}$, on a</u>

$$\forall (x,y) \in H_{\neq 1} \times H_{\neq 1} \quad : \quad h(x,y) = \frac{1}{(2\pi i)^n} Q(x,\bar{y}).$$

D'après [S] on peut par un changement analytique de coordonnées dans $\mathbb{C}^{n+1}$ au voisinage de zéro supposer que f est un polynôme tel que :

a) $Y(0) = \overline{f^{-1}(0)} \subset \mathbb{P}_{n+1}(\mathbb{C})$ a zéro pour unique point singulier.

b) $Y(t) = f^{-1}(t) \subset \mathbb{P}_{n+1}(\mathbb{C})$ est lisse pour $t \neq 0$, $|t|$ petit.

c) le degré de f est arbitrairement grand.

D'autre part il est montré dans (loc. cit. p. 296) que si le degré de f est suffisamment grand, la flèche de restriction $r_t : H^n(Y(t),\mathbb{C}) \to H^n(X(t),\mathbb{C})$ est surjective pour $t \neq 0$.

Notons $\underline{H}'$ le fibré sur Δ^* dont la fibre en t est $H^n(Y(t),\mathbb{C})$ et $H' = \Gamma(U,\pi^*\underline{H}')$.

On a donc que si le degré f est suffisamment grand la restriction $H' \xrightarrow{r} H$ est surjective.

D'après le théorème des cycles invariants [St1] on a alors une suite exacte :
$0 \to I \to H' \xrightarrow{r} H \to 0$, où I est l'espace des cycles de H' invariants par la monodromie. Si l'on note H'_1 l'espace propre généralisé associé à la valeur propre 1 de la monodromie on a alors (en notant toujours r le morphisme de restriction) une suite exacte :

$$0 \to I \to H'_1 \xrightarrow{r} H_1 \to 0 .$$

On a alors le résultat suivant, implicite dans [L2], énoncé et démontré dans [L4] :

Théorème II.1.9 : Si f vérifie les conditions a) et b) et si le degré de f est assez grand,

pour tout $(x,y) \in H_1 \times H_1$, on a :

$$h(x,y) = \frac{1}{(2\pi i)^n} Q'(N'\tilde{x} , \bar{\tilde{y}})$$

pour tout $\tilde{x}$ (resp. $\tilde{y}$) de H'_1 vérifiant $r(\tilde{x}) = x$ (resp. $r(\tilde{y}) = y$), avec Q' la forme d'intersection sur H' et $\frac{2\pi}{i} N'$ le logarithme de la partie unipotente de la monodromie sur H' (et il existe toujours de tels $\tilde{x}$ et $\tilde{y}$).

Indiquons maintenant comment nous démontrons le Théorème II.1.4.

Si $\omega \in \Gamma(B_\varepsilon , \Omega^{n+1})$, on pose $\alpha(\omega) = \inf\{\alpha \ / \ u^\omega_\alpha \neq 0\}$

et $w(\omega) = \inf\{k \ / \ u^\omega_{\alpha(\omega)} \in W_k H\}$, la filtration par le poids de la monodromie W_k étant normalisée par

$$W_n H = H , W_{-n} H = \{0\} .$$

Une réduction facile ramène la démonstration de II.1.4 à celle de la proposition suivante :

Proposition II.1.10 [L2] :

Soit $\omega \in \Gamma(B_\varepsilon , \Omega^{n+1})$ telle que :

(*) $\forall \omega' \in \Gamma(B_\varepsilon , \Omega^{n+1})$, si $\alpha(\omega') - \alpha(\omega) \in \mathbb{Z}$ alors $\alpha(\omega') \geq \alpha(\omega)$, alors l'image de $\int_{X(t)} \frac{\omega}{df} \wedge \frac{\bar{\omega}}{df}$ dans $\mathcal{P}$ est de la forme $K\,|t|^{2\alpha(\omega)} (\log|t|)^{w(\omega)} + o(|t|^{2\alpha(\omega)} \log|t|^{w(\omega)})$ avec $K \neq 0$, $w(\omega) \geq 0$ et $w(\omega) \geq 1$ si $\alpha(\omega) \in \mathbb{N}$.

Les inégalités sur $w(\omega)$ sont des conséquences formelles de la théorie de Hodge et sont démontrées dans [L2]. Pour démontrer la proposition il suffit, d'après

II.1.8 et II.1.9, de montrer que :

- $h(N^{w(\omega)}\ u^{\omega}_{\alpha(\omega)}\ ,\ u^{\omega}_{\alpha(\omega)}) \neq 0$ si $\alpha(\omega) \notin \mathbb{N}$
- $h(N^{w(\omega)-1}\ u^{\omega}_{\alpha(\omega)}\ ,\ u^{\omega}_{\alpha(\omega)}) \neq 0$ si $\alpha(\omega) \in \mathbb{N}$

Nous démontrons ceci en utilisant le Théorème de l'orbite $S\ell(2)$ de Schmid ([Sc]). En effet un corollaire du Théorème de Schmid est que les gradués par le poids d'une limite de structures de Hodge polarisées sont naturellement polarisés. Or Varchenko ([V2],[V3]) définit en utilisant les $\alpha(\omega)$ une filtration sur H qui coincide sur les gradués par le poids avec la filtration de Hodge limite de Steenbrink, qui, d'après ce qui précède, peut se déduire par restriction de la filtration de Hodge limite sur la cohomologie de la fibre de Milnor compactifiée, qui est justiciable du Théorème de Schmid.

Le Théorème II.1.3 dans le cas quasihomogène est une conséquence immédiate de II.1.4 car dans ce cas $E \cup -\mathbb{N}^* = B$. Le cas des courbes s'obtient en utilisant les résultats de Malgrange ([Ma]) qui relient le polynôme de Bernstein au saturé du système de Gauss-Manin et des arguments similaires à ceux que l'on vient d'exposer. Remarquons que le raisonnement de [L2] donne en général que si s est une racine de $\tilde{b}$, $s-n$ est un pôle de $|f|^{2s}$.

II.2 Une estimation géométrique du plus grand pôle non trivial de $|f|^{2s}$

Un autre problème qui se pose naturellement est d'exprimer les pôles de $|f|^{2s}$ en fonction d'invariants géométriques calculables à partir de f .

Le premier résultat obtenu dans cette direction est dû à J. Igusa :

Théorème II.2.1 [I2] :

Si $f : (\mathbb{C}^2,0) \to (\mathbb{C},0)$ définit un germe de courbe irréductible alors le plus grand pôle de $|f|^{2s}$ est :

$\frac{1}{m} + \frac{1}{\beta_1}$ où m est la multiplicité de f en zéro et β_1/m le premier exposant de Puiseux.

Si l'on remarque que $\beta_1 = 1 + \inf(\frac{e_q}{m_q})$ où les $\frac{e_q}{m_q}$ sont les invariants polaires de Teissier (d'après Merle [Me]), on peut poser la question d'exprimer le plus grand pôle non trivial de $|f|^{2s}$ (un pôle trivial est un pôle entier simple) en fonction d'invariants polaires.

Nous allons rappeler la définition des invariants polaires de Teissier dans le cas des singularités isolées d'hpersurfaces ([T1]).

Soient $x_o,\dots,x_n$ des coordonnées locales sur $\mathbb{C}^{n+1}$. Quitte à faire un changement linéaire de coordonnées, on peut supposer que les coordonnées locales sur $\mathbb{C}^{n+1}$ sont telles que l'hyperplan défini par $x_o = 0$ n'est pas limite en zéro d'hyperplans tangents à $f=0$.

Dans ce cas la courbe polaire Γ est définie par les équations $\frac{\partial f}{\partial x_1} = \frac{\partial f}{\partial x_2} = \dots = \frac{\partial f}{\partial x_n} = 0$, et si $\Gamma = \cup \Gamma_q$ est la décomposition de Γ en composantes irréductibles locales, on note m_q la multiplicité en zéro de Γ_q et e_q la multiplicité d'intersection en zéro de Γ_q avec $\frac{\partial f}{\partial x_o} = 0$. On définit alors

$$\theta^o(f) = \sup \left(\frac{e_q}{m_q}\right) \quad , \quad \tau^o(f) = \inf \left(\frac{e_q}{m_q}\right)$$

et

$$\theta^i(f) = \theta^o(f/H^i) \quad , \quad \tau^i(f) = \tau^o(f/H^i)$$

pour H^i un plan général de codimension i passant par l'origine, $1 \leqslant i \leqslant n-1$, $\theta^n(f) = \tau^n(f) = \mathrm{mult}_o(f=0)-1$.

Nous avons obtenu le résultat suivant (conjecturé dans un énoncé voisin par B. Teissier dans [T2]) :

<u>Théorème II.2.2</u> [L1] : <u>Soit</u> f : $(\mathbb{C}^{n+1},0) \to (\mathbb{C},0)$ <u>à singularité isolée, on a l'estimation suivante pour</u> $-\sigma(f)$, <u>le plus grand pôle non trivial de</u> $|f|^{2s}$:

$$\sum_{i=o}^{n} \frac{1}{1+[\tau^i(f)]} \leqslant \sigma(f) \leqslant \sum_{i=o}^{n} \frac{1}{1+\{\theta^i f\}} \quad ,$$

[ ] <u>désignant la partie entière et</u> { } = -[-]

Pour démontrer ce théorème on remarque que d'après II.1.5 $\sigma(f)$ coïncide avec l'exposant d'Arnold de f . On obtient alors le théorème par un argument de déformation et en utilisant que l'exposant d'Arnold ne varie pas dans une déformation à μ constant (d'après Varchenko [V4]) et est semi-continu dans une déformation arbitraire (d'après Steenbrink [St2]).

II.3 Quelques conséquences arithmétiques de la Théorie de Hodge

Comme les pôles de $|f|^s$ peuvent souvent s'exprimer en fonction d'invariants d'une résolution plongée de $f=0$, tant dans le cas complexe que dans le cas p-adique nous avons pu déduire de nos résultats archimédiens certaines conséquences arithmétiques :

Soit p un nombre premier, $\mathbb{Q}_p$ le corps des nombres p-adiques, k une extension finie de $\mathbb{Q}_p$ d'anneau des entiers R , P l'idéal maximal de R et $q = \mathrm{card}(R/P)$.

Soit f un polynôme de $R[x_1,\ldots,x_n]$ tel que $f(0) = 0$ qui a une singularité isolée en zéro. Quitte à faire une homothétie on peut supposer que f a zéro pour unique singularité dans la boule unité de $\overline{k}^n$ ($\overline{k}$ est la clôture algébrique de k).

Soit $\widetilde{N}_i = \mathrm{Card}\,\{x \bmod P^i \;/\; x \in R \text{ et } f(x) = 0 \bmod P^i\}$.
Alors nous avons les résultats suivants :

<u>Théorème II.3.1</u> [L5] : <u>Si la singularité de</u> f <u>en zéro n'est pas une singularité rationnelle*, on a</u> :

$$\limsup\, (\widetilde{N}_i)^{1/i} \leq q^{\left(n - \sum\limits_{i=o}^{n-1} \frac{1}{1+\{\theta^i(f)\}}\right)}$$

<u>Théorème II.3.2</u> [L5] : <u>Si</u> $f = 0$ <u>possède un point lisse k-rationnel dans</u> R^n <u>et si sur toute extension finie de</u> k <u>la contribution de la partie lisse de</u> $f = 0$ <u>à</u> $\widetilde{N}_i$ <u>est prépondérante, alors la singularité de</u> f <u>en zéro est nécessairement une singularité rationnelle</u>*.

La méthode que nous employons pour passer de $\mathbb{C}$ à $\mathbb{Q}_p$ est ad hoc et détournée. Cependant les exemples connus ([D2], [I2], [Li1], [Li2], [Li-M], [M2], [Str], [V1]) mettent en évidence une analogie frappante entre les pôles de $\int_{R^n} |f|^s\,|dx|$ (cas p-adique) et ceux de $|f|^{2s}$ (cas complexe).

Parmi les questions qui se posent naturellement au vu des exemples il y a les suivantes :

<u>Question II.3.3</u> [Li3] : Soit K une extension finie de $\mathbb{Q}$, $f \in K[x_1,\ldots,x_n]$ tel que $f(0) = 0$. Choisissant un plongement $\sigma : K \to \mathbb{C}$ on peut associer à f le polynôme ${}^{\sigma}f : \mathbb{C}^n \to \mathbb{C}$ dont on supposera que zéro est la seule valeur critique.

Soit p une place finie de K , K_p le complété de K en p d'anneau des entiers R_p .

Soit α un pôle réel non trivial de $\int_{R_p^n} |f(x)|_p^s\,|dx|_p$, $|\;|_p$ désignant la norme usuelle sur K_p^n .

Existe-t-il un point x de ${}^{\sigma}f^{-1}(0)$, un entier i , $1 \leq i \leq n-1$, et une valeur

* rappelons qu'une singularité rationnelle est une singularité de genre géométrique nul (cf [L5]).

propre ρ de la monodromie agissant sur $\mathcal{H}^i(R\,\psi({}^\sigma f))_x$, la fibre en x du i-ème faisceau de cohomologie du complexe des cycles évanescents le long de ${}^\sigma f^{-1}(0)$ (cf. [De]) tels que $\rho = e^{-2i\pi\alpha}$?

De plus on pourrait imaginer un lien entre la taille des blocs de Jordan associés à ρ et la multiplicité de α comme pôle.

Question II.3.4 (cf. [L3]) : Soit K une extension finie de $\mathbb{Q}_p$ d'anneau des entiers R et $f \in K[x_1,\ldots,x_n]$ tel que $f(0) = 0$. Soit b le polynôme de Bernstein algébrique de f et $\tilde{b}(s) = b(s)/s$ le polynôme de Bernstein réduit. Si α est un pôle réel non trivial de $\int_{R^n} |f(x)|^s\,|dx|$ existe-t-il une racine β de $\tilde{b}$ telle que $\beta-\alpha$ soit entier ? Si on est optimiste on peut demander si il existe une racine β telle que $\beta-\alpha$ soit strictement positif, ou même égal à 1, et s'il existe un lien entre la multiplicité de β comme racine et celle de α comme pôle.

III - POLES DE $\int d(x,V)^s$ ET VOLUMES DE TUBES

Soit V un sous ensemble analytique de $\mathbb{C}^N$. L'analogue de la fonction $I(s)$ de la première partie est l'intégrale

$$Z(s) = \int_P d(x,V)^s\,|dx|$$

où P est un polydisque compact dont on supposera le bord transverse à V, $d(x,V) = \inf\{d(x,y)/y \in V\}$, d étant la distance standard $d(x,y) = (\sum_{i=1}^n |x_i-y_i|^2)^{1/2}$ sur $\mathbb{C}^N$ et $|dx|$ la mesure de Lebesgue sur $\mathbb{C}^N$.

Théorème III.1 [L6]

La fonction $Z(s) = \int_P d(x,V)^s\,|dx|$ définie pour $\mathrm{Re}\, s > 0$ se prolonge analytiquement en une fonction méromorphe sur $\mathbb{C}$. Les pôles sont contenus dans un ensemble de la forme $-F-\mathbb{N}$, F étant un ensemble fini de rationnels positifs. L'ordre des pôles est au plus $2N+1$.

Si V est équidimensionnel de codimension d, le plus grand pôle de $Z(s)$ est $-2d$, c'est un pôle simple et le résidu est $\frac{2d\pi^d}{d!}\,\mathrm{vol}(V\cap P)$.

Comme la fonction $d(x,V)$ est seulement sous-analytique, le théorème III.1 ne se déduit pas des résultats classiques. Il nous faut utiliser les travaux d'Hironaka sur les ensembles sous-analytiques, en particulier son théorème de rectilinéarisation locale des ensembles sous analytiques ([H]).

Remarquons que dans le cas où V est algébrique on a un énoncé un peu meilleur que celui du théorème: les pôles de $Z(s)$ sont d'ordre au plus $2N$. En effet dans ce cas on écrit $\int_P d(x,V)^s \, |dx| = \int_\Gamma \varepsilon^s \, |dx|$ où $\Gamma \subset P \times \mathbb{R}$ est le graphe de $d(x,V)$ et les coordonnées sur $P \times \mathbb{R}$ sont (x,ε). Comme Γ est semi-algébrique réel d'après le Théorème de Tarski-Seidenberg, il est inclus dans une hypersurface algébrique réelle (de dimension réelle $2N$) et il est facile d'en déduire que dans ce cas l'ordre des pôles est au plus $2N$. Cet argument ne peut fonctionner dans le cas où V est analytique, car alors Γ est seulement sous analytique et ne peut être contenue dans une hypersurface analytique sans être nécessairement semi analytique. Cependant il est très probable que dans le cas analytique l'ordre des pôles est toujours au plus $2N$.

Dans le cas p-adique J. Denef a montré que l'intégrale analogue $I(s)$ (définie dans la première partie) se prolonge analytiquement en une fonction rationnelle de p^{-s} en utilisant le théorème de Macintyre, analogue p-adique du théorème de Tarski-Seidenberg ([D1], [D3], [D4]). Pour traiter le cas analytique p-adique il faudrait donc une bonne théorie des ensembles sous analytiques p-adiques ! D'autre part J. Oesterlé dans [O] a montré que dans le cas p-adique le plus grand pôle de $I(s)$ est $-2d$, que le pôle est simple et que le résidu est à une constante près le volume de $V \cap \mathbb{Z}_p^m$.

Nous allons maintenant faire le lien entre $\int_P d(x,V)^s$ et le volume des tubes à la Hermann-Weyl.

Soit $T(V,\varepsilon) = \{x \in \mathbb{C}^N \ / \ d(x,V) \leqslant \varepsilon\}$

et $T(V,\varepsilon) = \{x \in \mathbb{C}^N \ / \ d(x,V) = \varepsilon\}$.

On a la formule :

$$\int_P d(x,V)^s \, |dx| = \int_o^{+\infty} \varepsilon^s \, \mathrm{vol}(\partial T(V,\varepsilon) \cap P) \, d\varepsilon$$

Comme quand P est un polydisque compact ne rencontrant pas V , $\int_P d(x,V)^s \, |dx|$ est une fonction entière sur $\mathbb{C}$, on peut se localiser près de V pour étudier les pôles de $Z(s)$ et ses résidus. Ainsi $\varepsilon_o > 0$ étant fixé on pourra supposer que P est contenu dans le tube $T(V,\varepsilon_o)$.

Il est alors clair que III.1 équivaut au résultat suivant :

<u>Théorème III.1'</u> [L6] :

$J(\varepsilon) = \mathrm{vol}(T(V,\varepsilon) \cap P)$ <u>admet un développement asymptotique</u>

$$J(\varepsilon) = \sum_{\substack{\alpha \in -(F+\mathbb{N}) \\ 0 \leq k \leq 2N}} C_{\alpha,k} \, \varepsilon^{\alpha} (\log \varepsilon)^k \qquad (*)$$

<u>avec</u> F <u>un ensemble fini de rationnels positifs. De plus</u>

$J(\varepsilon) = \frac{\pi^d}{d!} \mathrm{vol}(V \cap P)\, \varepsilon^{2d} + o(\varepsilon^{2d})$ <u>si</u> d <u>est la codimension de</u> V (<u>supposé équidimensionnel</u>). <u>Si</u> V <u>est algébrique on peut écrire</u> $k \leq 2N-1$ <u>dans (*)</u>.

Rappelons que dans le cas où V est lisse, la formule d'Hermann-Weyl donne que pour ε petit $J(\varepsilon)$ est un polynôme en ε^2 dont les coefficients sont des intégrales de formes de Chern-Weil sur $V \cap P$ ([W], [Gr]).

Dans le cas où V a des singularités, le tube $T(V,\varepsilon)$ a toujours de l'autointersection, même pour ε petit :

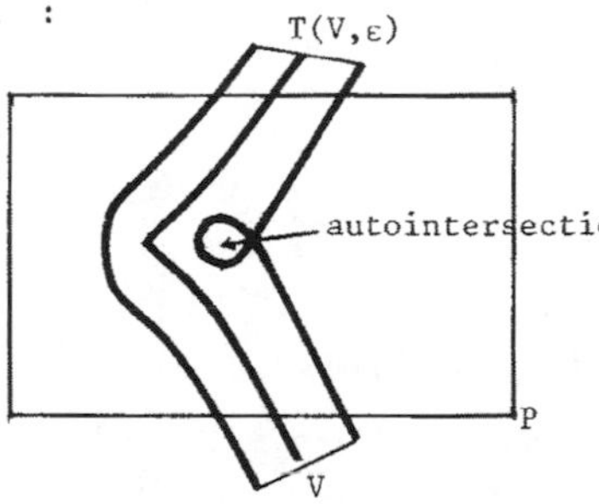

D'après le théorème III.1', dans le cas singulier on conserve un analogue de la formule d'Hermann-Weyl si on accepte des exposants rationnels et des logarithmes. Comme les exposants rationnels non entiers sont créés par les singularités de V il est naturel d'essayer de les relier à la géométrie des singularités de V.

C'est ce que nous sommes arrivés à faire pour le plus petit exposant non entier dans le cas des courbes planes, en l'exprimant comme un invariant <u>topologique</u> de la singularité.

Soit $V \subset \mathbb{C}^2$ une courbe plane ayant une singularité isolée à l'origine, et $P = D_\eta \times D_\eta$ un polydisque contenant l'origine comme unique singularité de V dont le bord est transverse à V, D_η désignant le disque fermé de centre zéro et rayon η.

On choisit (x,y) des coordonnées linéaires sur $\mathbb{C}^2$ de manière à ce que la projection sur l'axe des x soit finie.

Soit $f \in \mathbb{C}\{x,y\}$ l'équation de V, pour $y_\alpha(x)$ et $y_\beta(x)$ deux racines distinctes de f, on a $y_\alpha(x) - y_\beta(x) = C_{\alpha,\beta}\, x^{\kappa_{\alpha\beta}} + o(|x|^{\kappa_{\alpha\beta}})$ avec $C_{\alpha\beta} \neq 0$.

Suivant Lê Văn Thành ([Lê1], [Lê2]) on appelle $\kappa^x(V) = \max_{\alpha,\beta} \kappa_{\alpha,\beta}$ l'exposant de bifurcation maximale relatif à x de V, et $\kappa(V) = \sup_x \kappa^x(V)$ l'exposant de bifurcation maximal de V.

Dans [Lê1] et [Lê2], Lê Van Thành a démontré les propriétés suivantes de $\kappa(V)$

1) $\kappa(V) = \kappa^x(V)$ si x est un paramètre transverse.

2) On note $V = \bigcup_{1 \leqslant i \leqslant r} V_i$ la décomposition de V en composantes analytiquement irréductibles, β_s^i/m_i $(1 \leqslant s \leqslant g_i)$ les exposants caractéristiques de Puiseux de la branche V_i avec m_i la multiplicité de V_i. On dit que V_i et V_j sont équivalentes si et seulement si leurs développements de Puiseux coïncident jusqu'au dernier exposant caractéristique inclus. On note $\{\mathcal{X}_k\}_{1 \leqslant k \leqslant n}$ les classes d'équivalence des branches. Alors

$$\kappa(V) = \max_{1 \leqslant k \leqslant n} \left\{ \frac{\beta^k_{g_k}}{m_k} , \frac{1}{m_k}\left[\max_{V_i, V_j \in \mathcal{X}_k} (V_i, V_j) - \sum_{s=1}^{g_k} (d^k_s - d^k_{s+1})\, \beta^k_s \right] \right\}$$

avec $d^k_1 = m_k$, $d^k_{s+1} = \mathrm{pgcd}(m_k, \beta^k_1, \dots, \beta^k_s)$.

3) $\kappa(V) = \max \left\{ \frac{\beta^k_{g_k}}{m_k} , \frac{\max_{V_i, V_j \in \mathcal{X}_k} (V_i, V_j) - \mu_k + 1}{m_k} - 1 \right\}$, μ_k désignant le nombre de Milnor des branches de la classe $\mathcal{X}_k$.

On a le résultat suivant :

<u>Théorème III.2</u> [L6] : <u>Sous les hypothèses précédentes il existe</u> $C > 0$ <u>tel que</u> :

$$J(\varepsilon) = \pi \operatorname{vol}(V \cap P)\, \varepsilon^2 - C\, \varepsilon^{2(1+\kappa(V)^{-1})} + o(\varepsilon^{2(1+\kappa(V)^{-1})})$$

<u>Corollaire III.3</u> : <u>Sous les hypothèses précédentes le plus grand pôle de</u> $Z(s)$ <u>strictement inférieur à</u> -2 <u>est</u> $-2(1+\kappa(V)^{-1})$. <u>C'est un pôle simple de résidu strictement négatif.</u>

Pour donner une idée de la démonstration du théorème III.2 il est utile de définir "l'exposant d'injectivité" de V :

Soit $V \subset \mathbb{C}^N$ un sous ensemble analytique et P un polydisque compact dont le bord est transverse à V. On note SV le lien singulier de V.

On dit que le tube $T(r,\varepsilon) = T(V \cap P - T(SV,r),\varepsilon)$ n'a pas d'autointersection s'il est l'image difféomorphe par l'exponentielle du fibré normal à $V \cap P - T(SV,r)$ en disques de rayon ε .

Il est clair que pour tout $\varepsilon > 0$, il existe $r > 0$ tel que $T(r,\varepsilon)$ n'a pas d'autointersection. Par définition on note $r(\varepsilon)$ la borne inférieure des r tels que $T(r,\varepsilon)$ n'a pas d'autointersection.

On a le lemme facile suivant :

Lemme III.4 : Il existe des constantes strictement positives K et ρ telles que

$$r(\varepsilon) = K\,\varepsilon^{\rho} + o(\varepsilon^{\rho})$$

Définition III.5 : Sous les hypothèses précédentes on appelle ρ l'exposant d'injectivité.

Dans le cas où V est une courbe plane présentant une singularité isolée à l'origine on a la formule suivante pour ρ :

Proposition III.6 [L6] : Si V est une courbe plane avec une singularité isolée en zéro et P contient zéro pour unique singularité de V

$$\rho = \kappa(V)^{-1}$$

D'autre part toujours pour V une courbe plane on a l'estimation essentielle suivante (minoration du volume de l'autointersection) :

Proposition III.7 [L6] : Sous les hypothèses précédentes il existe une constante K , $0 < K < n\pi^2$ telle que

$$\mathrm{vol}(T(V \cap B(r(\varepsilon)),\varepsilon)) \leqslant K\, r^2(\varepsilon)\, \varepsilon^2 , \qquad \text{pour } \varepsilon \text{ petit,}$$

n désignant la multiplicité de V en zéro.

Le théorème III.2 se déduit facilement de III.6 et III.7 et de la formule d'Hermann-Weyl usuelle.

Remarquons que nous ne savons pas démontrer l'analogue de III.7 en dimension plus grande, ce qui permettrait d'exprimer le plus petit exposant non entier apparaissant dans $J(\varepsilon)$ en fonction de ρ .

Dans le cas p-adique D. Bollaerts a obtenu dans [Bo] un résultat analogue au nôtre :

Théorème III.8 (D. Bollaerts [Bo]) :

Soit $f(x_1,x_2) \in \mathbb{Z}_p [[x_1,x_2]]$ une série convergente sur $\mathbb{Z}_p^2$, qui a $(0,0)$ pour unique singularité et est analytiquement irréductible alors les pôles de $I(s)$

<u>sont de la forme</u> $\frac{2i\pi}{\log p}k-(\alpha+1)$ <u>avec</u> k <u>entier et</u> α <u>égal à 0, 1 à l'inverse d'un exposant de Puiseux.</u>

<u>Remarque III.9</u> : En fait la démonstration de D. Bollaerts montre que $-(1+\frac{n}{\beta g})$ est toujours pôle. On notera l'analogie avec III.3 car pour les courbes irréductibles $\frac{n}{\beta g}=\kappa^{-1}$.

BIBLIOGRAPHIE

[B1] D. Barlet : Développement asymptotique des fonctions obtenues par intégration dans les fibres. Inventiones Math. 68, p. 129-174 (1982).

[B2] D. Barlet : Contribution effective de la monodromie aux développements asymptotiques. Ann. Scient. Ec. Norm. Sup. 17, p. 293-315 (1984).

[B3] D. Barlet : Contribution du cup-produit de la fibre de Milnor aux pôles de $|f|^s$. Annales Institut Fourier 34, 4, p. 75-107 (1984).

[B4] D. Barlet : Monodromie et pôles du prolongement méromorphe de $\int_X |f|^{2\lambda} \square$. Prépublication de l'Institut E. Cartan (Nancy), Avril 1984.

[B5] D. Barlet : Forme hermitienne canonique sur la cohomologie de la fibre de Milnor d'une hypersurface à singularité isolée. Invent. Math. 81, p.115-153 (1985).

[Be] I.N. Bernstein : The analytic continuation of generalized functions with respect to a parameter. Funk. Analiz i Ego Pril. Vol. 6, n° 4, p. 26-40 (1972).

[Bo] D. Bollaerts : On the Poincaré series associated to the p-adic points on a curve. Preprint.

[Da] R. Danset : Méthode du cercle adélique et principe de Hasse fin pour certains systèmes de formes. L'Enseignement Mathématique t. 31 (1985) p. 1-66.

[D1] J. Denef : The rationality of the Poincaré series associated to the p-adic points on a variety. Invent. Math. 77, p. 1-23 (1984).

[D2] J. Denef : Poles of p-adic complex powers and Newton polyhedra. A paraître au Groupe d'Etude d'Analyse Ultramétrique, Paris 1984-85.

[D3] J. Denef : On the evaluation of certain p-adic integrals. Seminaire de Théorie des Nombres Paris 1983-84. Progress in Math. Birkhaüser p. 25-47.

[D4] J. Denef : P-adic semi algebraic sets and cell decomposition. Preprint 1985.

[De] P. Deligne : Le formalisme des cycles évanescents. SGA 7 exposé 13. Springer Lecture Notes vol. 340.

[G] P. Griffiths : Complex differential and integral geometry and curvature integrals associated to singularities of complex analytic varieties. Duke Math. Journal vol. 45 n° 3 p. 427-512 (1978).

[H] H. Hironaka : Introduction to Real-Analytic sets and Real-Analytic maps. Quaderni dei Gruppi ... Istituto Matematico "L. Tonelli" dell' Universita di Pisa (1973).

[I1] J.I. Igusa : Complex powers and asymptotic expansions. I), II). J. Reine augew Math. 268/269, p. 110-130 (1974) et 278/279, p. 307-321, (1975).

[I2] J.I. Igusa : On the first terms of certain asymptotic expansions. Complex and Algebraic Geometry. Iwanami Shoten and Cambridge Univ. Press, 1977, p. 357-368.

[I3] J.I. Igusa : On a certain Poisson formula. Nagoya Math. J. vol. 53 (1974) p. 211-233.

[I4] J.I. Igusa : Some observations on higher degree characters. American Journal of Math. vol. 99, n° 2 (1977), p. 393-417.

[I5] J.I. Igusa : Lectures on forms of higher degree. Springer-Verlag (1978).

[I6] J.I. Igusa : Some results on p-adic complex powers. American Journal of Math. vol. 106, n° 5 (1984), p. 1013-1032.

[I7] J.I. Igusa : Complex powers of irreducible algebraic curves. Preprint.

[I8] J.I. Igusa : Complex power of a hypersurface with isolated singularity. Preprint.

[I9] J.I. Igusa : Some aspects of the arithmetic theory of polynomials. Preprint.

[I10] J.I. Igusa : On functional equations of complex powers. Preprint.

[I11] J.I. Igusa : On a certain class of prehomogeneous vector spaces. Preprint.

[Lê1] Lê Văn Thành : Le lemme fondamental de Nilsson dans le cas analytique local. Ann. Inst. Fourier vol. 32 n° 1 (1982) p. 29-37.

[Lê2] Lê Văn Thành : Le nombre de Milnor et l'exposant de bifurcation. C.R. Acad. Sc. Paris t. 295, Série I (1982) p. 265-268.

[Li1] B. Lichtin : Some formulae for poles of $|f(x,y)|^s$. American Journal of Math. vol. 107, n° 1 (1985) p. 139-161.

[Li2] B. Lichtin : Poles of $|f|^{2s}$, Roots of the B-function, and an application to/from Knot theory. Preprint (Mai 1984).

[Li3] B. Lichtin : On the behaviour of Igusa's local zeta function in towers of field extension. Preprint.

[Li-M] B. Lichtin - D. Meuser : Poles of a local zeta function and Newtow polygons. Compositio Math. 55 (1985) p. 313-332.

[L1] F. Loeser : Exposant d'Arnold et sections planes. C.R. Academie des Sciences Paris. Tome 298, Série I (1984) p. 485-488.

[L2] F. Loeser : Quelques conséquences locales de la théorie de Hodge. Annales de l'Institut Fourier 35 n° 1 (1985) p. 75-92.

[L3] F. Loeser : Fonctions $|f|^s$, Théorie de Hodge et polynômes de Bernstein-Sato. A paraître aux Actes de la Conférence de Géométrie Algébrique de la Rabida.

[L4] F. Loeser : A propos de la forme hermitienne canonique d'une singularité isolée. A paraître au Bulletin de la S.M.F.

[L5] F. Loeser : Une estimation asymptotique du nombre de solutions approchées d'une équation p-adique . A paraître aux Inventiones Mathematicae.

[L6] F. Loeser : Volume de tubes autour de singularités. A paraître au Duke Mathematical Journal.

[Ma] B. Malgrange : Le polynôme de Bernstein d'une singularité isolée. Springer Lecture Notes 459 (1975) p. 98-115.

[Me] M. Merle : Invariants polaires des courbes planes. Invent. Math. 41 (1977), p. 103-111.

[M1] D. Meuser : On the rationality of certain generating functions. Math. Annalen 256, p. 303 -310 (1981).

[M2] D. Meuser : On the poles of a local zeta function for curves. Invent. Math. 73, p. 445-465 (1983).

[M3] D. Meuser : The meromorphic continuation of a zeta function of Weil and Igusa type. A paraître aux Inventiones Math.

[O] J. Oesterlé : Réduction modulo p^n des sous-ensembles analytiques fermés de $\mathbb{Z}_p^N$. Invent. Math. 66 p. 325-341 (1982).

[S] J. Scherk : On the monodromy theorem for isolated hypersurface singularities. Inventiones Math. 58, p. 289-301 (1980).

[Sc] W. Schmid : Variation of Hodge structure : the singularities of the period mapping. Invent. Math. 22 (1973), p. 211-319.

[Se] J.P. Serre : Quelques applications du théorème de densité de Chebotarev Publ. Math. IHES 54 (1981) p. 123-202.

[St1] J.H.M. Steenbrink : Mixed Hodge structure on vanishing cohomology. Real and complex singularities. Oslo 1976, p. 525-563.

[St2] J.H.M. Steenbrink : Semicontinuity of the singularity spectrum. Invent. Math. 79 (1985) p. 557-565.

[Str] L. Strauss : Poles of a two variable p-adic complex power. Transactions of the AMS, vol. 278 n° 2, (1983) p. 481-493.

[T1] B. Teissier : Variétés polaires I. Invariants des singularités d'hypersurfaces. Invent. Math. 40, n° 3 (1977), p. 267-292.

[T2] B. Teissier : Polyèdre de Newton jacobien. Séminaire sur les singularités Publications Mathématiques Paris VII (1980).

[V1] A. Varchenko : Newton polyedra and estimation of oscillating integrals. Funct. Anal. Appl. 10 (1976), p. 175-196.

[V2] A. Varchenko : Asymptotic mixed Hodge structure on vanishing cohomology. Izv. Akad. Nauk. 45, 3 (1981), p. 540-591.

[V3] A. Varchenko : Asymptotics of holomorphic forms define mixed Hodge structure. Dokl. Akad. Nauk. 255, 5 (1980) p. 1035-1038.

[V4] A. Varchenko : The complex Exponent of a singularity does not change along the strata μ = const. Funct. Anal. Appl. 16, n° 1 (1982) p. 1-12.

[W] H. Weyl : On the volume of tubes. Amer. J. of Math., vol. 61 (1939) p. 461-472.

SUR LES STRUCTURES DE HODGE MIXTES ASSOCIÉES AUX CYCLES EVANESCENTS

V. Navarro Aznar

Dept. de Matemàtiques, ETSEIB, Universitat Politècnica de Catalunya, Diagonal, 647, Barcelona, Espagne

§ 1. L'objet de cette note est d'exposer quelques résultats qui prouvent la présence de structures de Hodge mixtes sur divers invariants homotopiques tels que les H^i ou les π_i de la fibre de Milnor d'une fonction $f: X \to \mathbb{D}$ sur un espace analytique.

Plus précisement, soit X un espace analytique complexe de dimension $N+1$, $\mathbb{D}$ un disque dans $\mathbb{C}$, $f: X \to \mathbb{D}$ une fonction analytique sur X non-constante et V un ouvert de Zariski d'un sous-espace de $Y = f^{-1}(0)$ algébrique et compact. Par exemple, les deux situations qu'on rencontre le plus souvent dans la pratique sont la situation locale où $V = \{x\}$, $x \in X$, et la situation globale où f est propre et $V = Y$.

Notons $\mathbb{R}\Psi(\mathbb{Z})$ le complexe des cycles proches et $\mathbb{R}\Phi(\mathbb{Z})$ le complexe des cycles évanescents ([3]) et soit T l'endomorphisme de monodromie.

Le résultat principal qu'on démontre (voir les §§ suivants et [15]), qui généralise les résultats antérieurs de Clemens ([2]) et Steenbrink ([17], [18], [19], voir aussi [20]) et qui est en accord avec les résultats correspondants de Deligne dans [6], est le suivant:

1.1. <u>Théorème</u>. Les groupes de cohomologie $H^k(V, \mathbb{R}\Psi(\mathbb{Z}))$ et $H^k(V, \mathbb{R}\Phi(\mathbb{Z}))$, $k \geq 0$, ont des structures de Hodge mixtes, canoniques et fonctorielles en f et V, telles que:

i) la suite exacte

$$\ldots \to H^k(V, \mathbb{Z}) \to H^k(V, \mathbb{R}\Psi(\mathbb{Z})) \to H^k(V, \mathbb{R}\Phi(\mathbb{Z})) \to \ldots$$

est une suite exacte de structures de Hodge mixtes,

ii) le cup-produit de $H^*(V, \mathbb{R}\Psi(\mathbb{Z}))$ est un morphisme de structures de Hodge mixtes, comme l'est aussi le cup-produit de

$H^*(V, \mathbb{R}\Phi(\mathbb{Z})[-1])$,

iii) la partie semisimple T_s de la monodromie est un morphisme de structures de Hodge mixtes, et ainsi les sous-espaces propres généralisés qui correspondent a la valeur propre 1: $H^k(V, \mathbb{R}\Psi(\mathbb{Q}))_1$ et $H^k(V, \mathbb{R}\Phi(\mathbb{Q}))_1$, sont des sous-structures de Hodge mixtes de $H^k(V, \mathbb{R}\Psi(\mathbb{Q}))$ et $H^k(V, \mathbb{R}\Phi(\mathbb{Q}))$, $k \geq 0$, respectivement,

iv) le logarithme de la partie unipotente de la monodromie

$$N : H^k(V, \mathbb{R}\Psi(\mathbb{Z})) \longrightarrow H^k(V, \mathbb{R}\Psi(\mathbb{Z}))$$

est un morphisme de structures de Hodge mixtes de type $(-1, -1)$, qui s'insère dans une suite exacte de structures de Hodge mixtes (cf. §§ 3 et 4)

$$\ldots \to H^k(V, \mathbb{R}j_*\mathbb{Q}_{X^*}) \to H^k(V, \mathbb{R}\Psi(\mathbb{Q}))_1 \xrightarrow{N} H^k(V, \mathbb{R}\Psi(\mathbb{Q}))_1(-1) \to \ldots,$$

v) les nombres de Hodge h^{pq} de $H^k(V, \mathbb{R}\Psi(\mathbb{C}))$ vérifient:
si $k \leq N$, $h^{pq} = 0$, pour $(p, q) \notin [0, k] \times [0, k]$, et
si $k > N$, $h^{pq} = 0$, pour $(p, q) \notin [k-N, N] \times [k-N, N]$,

vi) si $X-Y$ est non-singulier et x est une singularité rationnelle de X , la filtration de Hodge F vérifie

$$Gr_F^0 \, [\mathbb{R}^k\Psi(\mathbb{Q})_x]_1 = 0 \text{ , pour tout } k > 0 .$$

Il résulte de iv) et v):

1.2. <u>Corollaire</u>. (cf. [8], [14], [13], [16]). Soit l le plus grand nombre de composantes $Gr_F^p \, H^k(V, \mathbb{R}\Psi(\mathbb{C}))$ successives non-nulles (en particulier $l \leq N+1$) , alors

$$(T_s - 1)^l = 0 .$$

Et il résulte de iv), v) et vi):

1.3. <u>Corollaire</u>. (cf. [1], [18]). Avec les hypothèses de vi), les blocs de Jordan de la monodromie T sur $\mathbb{R}^k\Psi(\mathbb{Q})_x$, $k > 0$, qui correspondent à la valeur propre 1 sont de taille au plus k .

§ 2. Avant d'esquisser la preuve du théorème 1.1., on va introduire un formalisme de foncteurs derivés de Thom-Whitney qui nous sera utile par la suite et qui nous permettrait aussi de prouver la présence des structures de Hodge mixtes naturelles sur les groupes d'homotopie rationnelle, voir [15]. (Pendant la Conférence de Sant Cugat, j'ai appris que R. Hain a obtenu ce résultat sur les π_i , indépendamment et par une méthode différente, dans la situation globale etudiée par Clemens et Steenbrink, voir [12]).

Le problème qui résout ce formalisme est une version faisceautique du problème des cochaines commutatives formulé par Thom, et etudié indépendamment par Whitney. Rappelons que ce problème, résolu par Quillen et aussi par Sullivan, demandait la construction fonctorielle pour tout espace topologique X d'une $\mathbb{Q}$-algèbre dgc $A^*(X, \mathbb{Q})$ isomorphe dans $D^+(\mathbb{Q})$ à $C^*(X, \mathbb{Q})$, l'algèbre dg des $\mathbb{Q}$-cochaines singulières de X , par un isomorphisme qui induise un isomorphisme d'algèbres en cohomologie, et telle que si Y est un sous-espace fermé de X , $A^*(Y, \mathbb{Q})$ soit un quotient de $A^*(X, \mathbb{Q})$.

Soit $\underline{k}$ un corps de caractéristique 0 et X un espace topologique. Notons $C^+(X, \underline{k})$ la catégorie des complexes inférieurement bornés de faisceaux de $\underline{k}$-espaces vectoriels sur X et $A(X, \underline{k})$ la catégorie des faisceaux de k-algèbres dgc sur X .

On sait d'après la théorie classique de Godement que, si $f\colon X \to Y$ est une application continue entre des espaces topologiques, on a un foncteur

$$\mathbb{R}f_*\colon C^+(X, \underline{k}) \longrightarrow C^+(Y, \underline{k}) .$$

On démontre dans [15] le résultat suivant:

2.1. <u>Théorème</u>. Avec les notations précédentes. Il existe un foncteur

$$\mathbb{R}_{TW}f_*\colon A(X, \underline{k}) \longrightarrow A(Y, \underline{k})$$

et une transformation naturelle I des foncteurs

$$A(X, \underline{k}) \xrightarrow{\mathbb{R}_{TW}f_*} A(Y, \underline{k}) \longrightarrow C^+(Y, \underline{k})$$

$$A(X, \underline{k}) \longrightarrow C^+(X, \underline{k}) \xrightarrow{\mathbb{R}f_*} C^+(Y, \underline{k})$$

telle que

$$I_A : \mathbb{R}_{TW} f_*(A) \longrightarrow \mathbb{R} f_*(A)$$

est un quasi-isomorphisme qui induit un isomorphisme d'algèbres en cohomologie, pour tout $A \in \mathrm{Ob}\, A(X, \underline{k})$.

Par exemple, considérons sur un espace topologique X le faisceau constant $\underline{k}_X$, d'après 2.1. on obtient une k-algèbre dgc $\mathbb{R}_{TW}\Gamma(X, \underline{k}_X)$ qu'on peut considérer comme la solution faisceautique du problème des cochaines commutatives, et qui garde avec l'algèbre de Sullivan $Su^*(X, \underline{k})$ ([21], [22]) la même relation qu'il y a entre les groupes de cohomologie de X à valeurs dans $\underline{k}_X$: $H^*(X, \underline{k}_X)$, et la cohomologie singulière de X à coefficients dans $\underline{k}$: $H^*_{sing}(X, \underline{k})$. De fait, pour relier ces deux $\underline{k}$-algèbres, on introduit le faisceau Su^*_X qui résulte de faisceautiser les algèbres de Sullivan, et on démontre que si X est paracompacte et $\underline{k}$-HLC , alors Su^*_X est une résolution par des faisceaux mous de $\underline{k}_X$. Comme conséquence, on obtient que si X est paracompacte et $\underline{k}$-HLC , les $\underline{k}$-algèbres dgc $\mathbb{R}_{TW}\Gamma(X, \underline{k}_X)$ et $Su^*(X, \underline{k})$ sont isomorphes dans $\mathrm{Ho}\, A(\underline{k})$ (voir [15], § 5).

Nous désignerons dans ce qui suit par Su^*_X la résolution ainsi obtenue de $\mathbb{Q}_X$.

§ 3. Esquissons maintenant la preuve de 1.1.

D'après les résultats de [10] (voir aussi [9], [11]) il existe un schéma simplicial strict augmenté

$$\pi_\bullet : X_\bullet \longrightarrow X$$

qui est de descente cohomologique sur $X-Y$, et tel que, pour tout $p \geq 0$, on a:

i) X_p est non-singulier,

ii) les morphismes $\pi_p : X_p \to X$ sont projectifs,

iii) la fibre $Y_p = (f \circ \pi_p)^{-1}(0)$ est un diviseur à croisements normaux dans X_p , dont toutes les composantes irréductibles sont non-singulières,

iv) $\bar{V}_p = \pi_p^{-1}(\bar{V})$ et $W_p = \pi_p^{-1}(\bar{V}-V)$ sont des réunions de composantes irréductibles de Y_p,

v) $\dim X_p \leq \dim X-p$.

On démontre aussi dans loc. cit. que la catégorie de telles hyperrésolutions vérifiant i)-iv) est connexe.

Il en résulte, par la théorie de la descente cohomologique de Deligne et les résultats de Deligne sur les complexes de Hodge simpliciaux ([5]), que dans la preuve de 1.1. on peut supposer que:

i) X est non-singulier,

ii) Y est un diviseur à croisements normaux dans X dont toutes les composantes sont non-singulières, et

iii) V est une réunion de composantes irréductibles de Y .

En utilisant une décomposition de Mayer-Vietoris pour $\mathbb{R}\Gamma(V, \mathbb{R}\psi)$, on voit qu'il suffit de prouver le théorème 1.1. en supposant au lieu de iii):

iii bis) V est une intersection de composantes de Y , i.e. si on pose $Y = Y_1 \cup Y_2 \cup \dots \cup Y_r$, alors

$$V = Y_{i_1} \cap Y_{i_2} \cap \dots \cap Y_{i_s} \quad (=:Y_I) ,$$

pour un ensemble d'index

$$I = (i_1, i_2, \dots, i_s) , \ 1 \leq i_1 < i_2 < \dots < i_s \leq r .$$

Soit $H^*(V, \mathbb{R}\psi(\mathbb{Q}))_1$ le sous-espace propre généralisé qui correspond à la valeur propre 1 de la monodromie, on va construire un complexe de Hodge mixte cohomologique sur V dont l'hypercohomologie sera $H^*(V, \mathbb{R}\psi(\mathbb{Q}))_1$. Ce complexe munira $H^*(V, \mathbb{R}\psi(\mathbb{Q}))$ d'une structure de Hodge mixte si la monodromie est unipotente, et, en général, il suffira de faire un changement de base finie.

Considérons d'abord le niveau sur $\mathbb{C}$. D'après [3], (voir aussi [17]) on a

$$H^*(V, \mathbb{R}\psi(\mathbb{Q}))_1 \cong H^*(V, \Omega_X^*(\log Y)[\log t]) ,$$

et en oubliant pour l'instant le log t , on trouve que le problème sur lequel on retombe est celui de localiser la construction du complexe logarithmique ([4]) à une intersection V de composantes irréductibles de Y .

Notons $\underline{J}_I$ l'idéal de $V(=Y_I)$ dans $\underline{O}_X$ et $d\underline{J}_I$ le sous-module de Ω_X^1 engendré par l'image de $\underline{J}_I$ dans Ω_X^1 par d .

Posons

$$\Omega_X^p(\log Y;\, Y_I) = \frac{\Omega_X^p(\log Y)}{dJ_I \wedge \Omega_X^{p-1}(\log Y) + J_I\Omega_X^p(\log Y)}\ , \quad p \geq 0\ ,$$

avec la différentielle induite par celle de $\Omega_X^*(\log Y)$, $\Omega_X^*(\log Y;\, Y_I)$ est une $\mathbb{C}$-algèbre dgc.

On définit la filtration par le poids sur $\Omega_X^*(\log Y; Y_I)$ par

$$W_k\Omega_X^p(\log Y;\, Y_I) = \frac{W_k\Omega_X^p(\log Y)}{dJ_I \wedge W_k\Omega_X^{p-1}(\log Y) + J_I W_k\Omega_X^p(\log Y)}\ , \quad p,k \geq 0,$$

qui est une filtration croissante et multiplicative.

On définit la filtration de Hodge sur $\Omega_X^*(\log Y;\, Y_I)$ par

$$F^l\Omega_X^*(\log Y;\, Y_I) = \bigoplus_{p\geq l} \Omega_X^p(\log Y;\, Y_I)\ , \quad l \geq 0\ ,$$

c'est à dire la filtration bête de $\Omega_X^*(\log Y;\, Y_I)$, qui est une filtration décroissante et multiplicative.

Notons

$$\tilde{Y}^{(k)} = \coprod_{|J|=k} Y_J$$

et

$$\tilde{Y}(I)^{(k)} = \coprod_{|J|=k} Y_I \cap Y_J\ .$$

Il est immédiat que le résidu de Poincaré

$$\text{Rés}_k\colon\ W_k\Omega_X^p(\log Y) \longrightarrow \Omega^{p-k}_{\tilde{Y}^{(k)}}\ , \quad k,p \geq 0\ ,$$

passe aux quotients et induit des morphismes

$$\text{Rés}_k\colon\ Gr_k^W\Omega_X^p(\log Y;\, Y_I) \longrightarrow \Omega^{p-k}_{\tilde{Y}(I)^{(k)}}\ , \quad k,p \geq 0\ ,$$

compatibles avec la différentielle, et qui sont de fait des isomorphismes, comme il résulte d'un calcul en coordonnées locales.

Notons $U = X-Y$ et soit $j\colon U \to X$ l'inclusion.

Il est maintenant aisé, en suivant de près les arguments de

Deligne dans [4], de montrer que

$$(\mathbb{R}j_*\mathbb{Z}_{U|V},\ (j_*Su^*_{U|V},\ \tau),\ (\Omega^*_X(\log Y;\ V),\ W,\ F))$$

est un complexe de Hodge mixte cohomologique sur V, qui munit l'algèbre $H^*(V,\ \mathbb{R}j_*Z_U)$ d'une structure de Hodge mixte, fonctorielle en X,Y et V (voir [15], §10).

Arrivés à ce point, on constate qu'en cherchant à prouver le théorème 1.1., on a prouvé d'abord le théorème suivant (voir [15], § 13).

3.1. Théorème. Soit X un espace analytique complexe, Y un sous-espace analytique fermé de X et V un ouvert de Zariski d'un sous-espace de Y algébrique et compact. Posons $U = X-Y$ et soit $j: U \to X$ l'inclusion, alors les algèbres $H^*(V,\ \mathbb{R}j_*\mathbb{Z}_U)$ et $H^*(V,\ \mathbb{R}\Gamma_Y\mathbb{Z}_X)$ ont des structures de Hodge mixtes, canoniques et fonctorielles en X,Y et V, telles que:

i) la suite exacte

$$\ldots \to H^k(V,\ \mathbb{R}\Gamma_Y\mathbb{Z}_X) \to H^k(V,\ \mathbb{Z}) \to H^k(V,\ \mathbb{R}j_*\mathbb{Z}_U) \to \ldots$$

est une suite exacte de structures de Hodge mixtes,

ii) le cup-produit

$$H^*(V,\ \mathbb{Z}) \otimes H^*(V,\ \mathbb{R}\Gamma_Y\mathbb{Z}_X) \longrightarrow H^*(V,\ \mathbb{R}\Gamma_Y\mathbb{Z}_X)$$

est un morphisme de structures de Hodge mixtes.

Par exemple, si X est un espace analytique et x est un point de X, puisque la cohomologie de $(\mathbb{R}j_*\mathbb{Z}_{X-\{x\}})_x$ est la cohomologie du noeud de X en x, on trouve d'après 3.1. une structure de Hodge mixte sur cette cohomologie.

(Pendant la Conférence de Sant Cugat, j'ai appris que A. Durfee et R. Hain ont obtenu indépendamment et avec une construction distincte, des résultats proches du théorème 3.1., voir [7]).

Revenons à la preuve de 1.1., en continuant sur le niveau $\mathbb{C}$.

Sur le complexe $\Omega^*_X(\log Y;\ V)[\log t]$, on définit la filtration par le poids M comme la convolution de la filtration W de $\Omega^*_X(\log Y;\ V)$ avec la filtration croissante qui donne à $\log t$ le poids 2, et on définit la filtration de Hodge F comme la convolution de la filtration de Hodge de $\Omega^*_X(\log Y;\, V)$ avec la filtration décroissante par le degré en $\log t$. Il en résulte que $(\Omega^*_X(\log Y;\, V)[\log t],\, M,\, F)$ est un faisceau de $\mathbb{C}$-algèbres dgc bifiltré, quasi-isomorphe à $\mathbb{R}\psi(\mathbb{C})_1$.

Il nous reste maintenant à traduire la construction précédente au niveau rationnel, tâche pour laquelle le formalisme du § 2 nous devient très utile.

En effet, soit $\mathbb{D}^*$ le disque pointé $\mathbb{D}-\{0\}$, et $\tilde{\mathbb{D}}^*$ le demiplan de Poincaré $\{z \in \mathbb{C},\ \mathrm{Im}\, z > 0\}$. L'application $z \longrightarrow \exp(2\pi i z) = t$ fait de $\tilde{\mathbb{D}}^*$ un recouvrement universel de $\mathbb{D}^*$, donc si on pose $X^* = X \times_{\mathbb{D}} \mathbb{D}^*$, $\tilde{X}^* = X \times_{\mathbb{D}} \tilde{\mathbb{D}}^*$ et $j: X^* \longrightarrow X$, et $\tilde{j}: \tilde{X}^* \longrightarrow X$ sont les projections naturelles, on obtient le diagramme

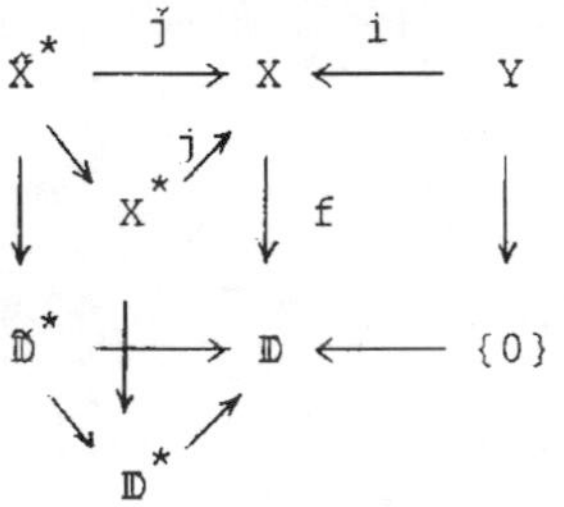

Notons $\iota: \mathbb{D}^* \longrightarrow \mathbb{D}$ l'inclusion naturelle, et soit θ une section fermée de $\iota_* Su^1_{\mathbb{D}^*}$ sur $\mathbb{D}$, qui représente un générateur positif du $H^1(\iota_* Su^*_{\mathbb{D}^*})_0$.

Puisque $H^1(\tilde{\mathbb{D}}^*, \mathbb{Q}) = 0$ et $d\theta = 0$ sur $\mathbb{D}^*$, on a $\pi^*\theta = d\eta$, avec η une section de $Su^0_{\tilde{\mathbb{D}}^*}$ sur $\tilde{\mathbb{D}}^*$. Par suite, l'image $f^*\eta$ de η par l'application

$$f^*: \Gamma(\tilde{\mathbb{D}}^*,\, Su^*_{\tilde{\mathbb{D}}^*}) \longrightarrow \Gamma(\tilde{X}^*,\, Su^*_{\tilde{X}^*})$$

définit une section, encore notée η , de $\tilde{j}_* Su^0_{\tilde{X}^*}$ sur X .

Puisque $j_* Su^*_{X^*}$ est une sous-algèbre de $\tilde{j}_* Su^*_{\tilde{X}^*}$ et $d\eta \in j_* Su^1_{X^*}$, l'extension de Hirsch $(j_* Su^*_{X^*})[\eta]$ est une sous-algèbre de $\tilde{j}_* Su^*_{\tilde{X}^*}$, stable par la monodromie, et qui, filtrée par la convolution M de la filtration qui donne à η le poids 2 avec la filtration canonique τ de $j_* Su^*_{X^*}$, définit un faisceau de $\mathbb{Q}$-algèbres dgc filtrées sur V .

On vérifie que les faisceaux $((j_* Su^*_{X^*})[n], M) \otimes_{\mathbb{Q}} \mathbb{C}$ et $(\Omega^*_X(\log Y;\ V), M)$ sont quasi-isomorphes filtrés, et on obtient finalement (voir [15], § 14) que

$$[((j_* Su^*_{X^*})[n], M),\ (\Omega^*_X(\log Y;\ V)[\log t], M, F)]$$

est un complexe de Hodge mixte cohomologique sur V qui munit l'algèbre $H^*(V, \mathbb{R}\psi(\mathbb{Q}))_1$ d'une structure de Hodge mixte canonique et fonctorielle, en f et V .

Comme il a été indiqué au début de ce §, cette construction permet, d'après la théorie de la descente cohomologique et les résultats de [5], de munir les groupes $H^k(V, \mathbb{R}\psi(\mathbb{Z}))$ et $H^k(V, \mathbb{R}\Phi(\mathbb{Z}))$, $k \geq 0$, V comme dans 1.1., de structures de Hodge mixtes, canoniques et fonctorielles. Nous renvoyons le lecteur à [15], pour la preuve des propriétés i)-vi).

§ 4. En conclusion, les théorèmes précédents 1.1. et 3.1. montrent que les fibres $(\mathbb{R}j_*\mathbb{Z}_U)_x$, $(\mathbb{R}\Gamma_Y\mathbb{Z}_X)_x$ et $\mathbb{R}\psi(\mathbb{Z})_x$ ont des structures de Hodge mixtes, résultat qu'on peut interpréter comme un indice de la traduction transcendente des faisceaux mixtes de [6]; ils montrent de même que les groupes de cohomologie de ces complexes sur une variété algébrique ont des structures de Hodge mixtes, résultat qu'on peut aussi interpréter comme un indice de la traduction transcendente de la stabilité des faisceaux mixtes par le foncteur images directes. On espère que cette traduction transcendente du formalisme des faisceaux mixtes pourra être un jour complètement découverte.

BIBLIOGRAPHIE

[1] D.Barlet: Contribution du cup-produit de la fibre de Milnor aux pôles de $|f|^2$, Ann. Inst. Fourier, Grenoble, 34, 4 (1984), 75-107.

[2] C.H.Clemens: Degenerations of Kähler manifolds, Duke Math. Journal, 44 (1977), 215-290.

[3] P.Deligne: Comparaison avec la théorie transcendente, Exp. XIV, dans SGA 7 II, Lecture Notes in Mathematics nº 340, 1973.

[4] P.Deligne: Théorie de Hodge II, Publ. Math. I.H.E.S., 40 (1972), 5-57.

[5] P.Deligne: Théorie de Hodge III, Publ. Math.I.H.E.S., 44 (1975), 5-77.

[6] P.Deligne: La conjecture de Weil, II, Publ. Math. I.H.E.S., 52 (1980), 137-252.

[7] A.Durfee-R.Hain: Mixed Hodge structures on the homotopy of links, prepublication.

[8] A.Grothendieck: Classes de Chern et représentations lineaires des groupes discrets, dans Dix exposés sur la cohomologie des schémas, North-Holland, 1968.

[9] F.Guillén: Une relation entre la filtration de Zeeman et la filtration par le poids de Deligne, prepublication, Universitat Politècnica de Catalunya, à paraitre dans Compositio Math.

[10] F.Guillén-V.Navarro Aznar-F.Puerta: Théorie de Hodge via schémas cubiques, notes policopiées, Universitat Politècnica de Catalunya, 1982.

[11] F.Guillén-F.Puerta: Hyperrésolutions cubiques et applications à la théorie de Hodge-Deligne, dans ce volume.

[12] R.Hain: Mixed Hodge structures on homotopy groups, Bull. Amer. Math. Soc. (New Series), 14, 111-114(1986).

[13] N.Katz: Nilpotent connections and the monodromy theorem: applications of a result of Turritin, Publ. Math. IHES, 39, 175-232 (1971).

[14] A.Landman: On the Picard-Lefschetz transformation for algebraic manifolds acquiring general singularities, Trans. Amer. Math. Soc., 181 (1973), 89-126.

[15] V.Navarro Aznar: Sur la théorie de Hodge-Deligne, prepublication, Universitat Politècnica de Catalunya.

[16] W.Schmid: Variation of Hodge structure: The singularities of the period mapping, Invent. math., 22 (1973), 21-320.

[17] J.H.Steenbrink: Limits of Hodge structures, Invent. math., 31 (1976), 229-257.

[18] J.H.M.Steenbrink: Mixed Hodge structure on the vanishing cohomology, Real and Complex Singularities (Oslo, 1976), Noordhoff-Sijthoff, 1977.

[19] J.H.M.Steenbrink: Mixed Hodge structures associated with isolated singularities, Proc. of Symp. in Pure Math., vol. 40 part 2, (1983), 513-536.

[20] J.H.Steenbrink - S.Zucker: Variation of mixed Hodge structure.I, Invent. math., 80 (1985), 489-542.

[21] D.Sullivan: Infinitesimal computations in topology, Publ.I.H.E.S., 47 (1977), 269-331.

[22] R.Swan: Thom's theory of differential forms on simplicial sets, Topology, vol 44 (1975), 271-273.

L^2-COHOMOLOGY OF ALGEBRAIC VARIETIES IN THE FUBINI METRIC

Vishwambhar Pati
Department of Mathematics
Harvard University
Cambridge, MA 02138

1. Introduction

It was conjectured by Cheeger-Goresky-MacPherson in [2] that the L^2-cohomology of the non-singular part of a projective variety X in $\mathbb{C}\mathbb{P}(N)$ with the induced (incomplete Fubini-metric (given by $\partial\bar{\partial} \log \sum_{i=0}^{N} |z_i|^2$, z_i homogeneous coordinates in $\mathbb{C}\mathbb{P}(N)$) is isomorphic to middle perversity intersection homology. We recall that for a singular space X with singular set Σ and metric g in the non-singular part $X-\Sigma$, the L^2-cohomology (of $X-\Sigma$), denoted by $H^i_{(2)}(X-\Sigma)$ is the cohomology of the subcomplex of the de-Rham complex defined by

$$(\text{dom } d_{X-\Sigma})^i = \{\alpha \in \Lambda^i(X-\Sigma) : \int_{X-\Sigma} \alpha \wedge * \alpha < \infty, \quad \int_{X-\Sigma} d\alpha \wedge *(d\alpha) < \infty\}$$

(where $*$ is with resepct to g). The middle-perversity intersection homology of a PL-stratified space $X = X^n \hookleftarrow X^{n-2} = \Sigma \hookleftarrow X^{n-4} \cdots \hookleftarrow X^{n-2k} \hookleftarrow \cdots X^{-1} = \phi$ (with even-codim. strata, say) is the homology of the subcomplex of the simplicial chain complex of X defined by

$$IC_i(X) = \{\sigma \in C_i(X): \dim(|\sigma| \cap X^{n-2k}) \leq i-k-1, \quad \dim(|\partial\sigma| \cap X^{n-2k}) \leq i-k-2\}.$$

(The coefficient groups are taken to be $\mathbb{R}$ or $\mathbb{C}$ throughout.) Since de-Rham cohomology and usual homology are particular cases of L^2-cohomology and intersection homology for non-singular varieties ($\Sigma = \phi$), the conjecture may be viewed as a natural generalization of the de-Rham Theorem to singular varieties. For further details about these notions, see [1], [2], and [3].

In [1], Cheeger shows that the conjecture is true for stratified spaces that are admissible Riemannian pseudomanifolds. For stratified pseudomanifolds with even-codim. strata, this basically boils down to saying that $X-\Sigma$ is built up of "admissible handles". An admissible handle is $B^{n-2k} \times c(N^{2k-1})$ (c denotes the deleted open cone) so that the metric, when restricted to it, is quasi-isometric to (usual Euclidean metric on ball) $\times (dr^2 + r^2 g_N)$ (where r is the cone parameter and g_N is a nonsingular metric on the "link" N, where N is itself admissible, inductively defined). In particular, his result implies the conjecture for projective curves, because the metric in the neighborhood of a singular point of an algebraic curve is always quasi-isometric to a conical metric, as can be verified by normalization.

In this article, we will sketch a proof of the conjecture for normal projective surfaces. This was established by the author, jointly with W-C. Hsiang in [4]. Actually, as we will see, in the case of surfaces, the assumption of normality can be dropped. In [7], it is also proved that the conjecture is true for a certain class of 3-folds with isolated singularities. The main technique is to pull back the induced metric to a sufficiently "high" resolution of singularities. This allows us to break the metric up into (quasi-isometric) simple models in small neighborhoods of the exceptional divisor. For these models, Cheeger-type L^2-estimates hold as in [1], and the estimates globalise by using a piecewise smooth vector field which is constructed essentially by hand.

It is hoped that the validity of the conjecture would afford an analytic approach to putting Hodge structures on $IH_*(X)$ for singular varieties X. For more discussion on this (as well as the "Kahler package") see [1], [2]. Indeed, most of the classical "Kahler package" for non-singular projective varieties has been shown to be true for $IH_*(X)$, if X is singular, by essentially non-analytic methods.

2. Main Steps

For the time being, let V be a normal projective surface, so that it has isolated singularities ($\Sigma = \{p_1,\ldots,p_\ell\}$). A small neighborhood of each p_i is topologically a cone on some (non-singular) link N, which is connected and compact. Also, in small affine neighborhoods of points in $\mathbb{CP}(N)$, the Fubini metric is quasi-isometric to the standard Euclidean metric $\sum_{i=1}^{N} dz_i d\bar{z}_i$ (z_i, affine coordinates). So if we are looking at a small neighborhood of a singular point, it is enough to assume that V is affine algebraic, with a singular point at the origin. Denote $\{V-0\} \cap B_\varepsilon(0)$ by $V_\varepsilon = cN \overset{\text{homeo}}{\sim} (0,1)\times N$. If one shows that

$$\text{(Poincaré Lemma)} \qquad H^i_{(2)}(V_\varepsilon) = H^i_{(2)}(N) \quad \text{for } i \leq 1$$
$$= 0 \qquad \text{for } i > 2$$

then one would be done, by a Mayer-Vietoris argument, see [1]. This is called the "Poincaré Lemma" in analogy with the non-singular case.

Of course, one needs to remark that this (local) duality with $IH_*(\overline{V}_\varepsilon)$ is induced by a global integration map, see [1]. (For the calculation of IH_* for a cone on N (<u>including</u> the vertex, as 0-stratum), see [3]). That the local calculations for $H^*_{(2)}$ and IH_* agree for a cone was observed by Sullivan and motivated the conjecture in question.

2.1 <u>Standard form of coordinate-functions</u>

So $V \subset \mathbb{C}^N$, with a singularity at 0, is an affine normal surface. Let $\pi\colon \tilde{V} \longrightarrow V$ be a resolution of singularities. Then $\pi^{-1}(0)$ is a union of curves $\bigcup_i D_i$ such that D_i meets D_j transversely, if at all. Also $\pi^{-1}(V_\varepsilon)$ is an ε-tubular neighborhood (with the core $\bigcup_i D_i$ deleted) which is analytically isomorphic to V_ε. Cover this neighborhood with (u,v)-coordinate patches such that $\bigcup_i D_i$ is locally defined by $(u = 0)$ (away from a normal crossing) or $(u=0) \cup (v=0)$ (at a normal crossing). We would like to get the original coordinates $z_1,\ldots,z_N$ on V in a good standard form in terms of u and v. At this point

$$z_i = u^{m_i} v^{n_i} h_i(u,v) \qquad (i = 1,\ldots,N)$$

where (u,v) is a holomorphic power series in u and v. This means that the coordinate hyperplane section $(z_i = 0)$ on V lifts to the (u,v) neighborhood as $(u=0) \cup (v=0) \cup (h_i=0)$, a union of curves. Now further planar blow-ups of the origin in the (u,v) chart can be done so that $z_i = 0$ lifts to non-singular curves which intersect only pairwise, if at all. Hence after enough blow ups, one may assume that in a small (u,v) neighborhood, $h_i \neq 0$, i.e., that h_i is a local unit, with non-zero constant term, and in any (u,v) chart $(z_i = 0)$ lifts to either $u = 0$, or $(u = 0,\ v = 0)$. Hence $(u = 0)$ will always blow-down to the origin in $\mathbb{C}^N$, and $v = 0$ may or may not depending on whether the chart sits on a normal crossing of $\bigcup_i D_i$ or not. Hence

$$z_i = u^{n_i} v^{m_i} \text{ (local unit)} \qquad (i = 1,\ldots,N)$$

after enough blow-ups. After enough blow-ups, it may also be arranged that $n_1 \leq n_i$, $m_1 \leq m_i$ $\forall i$. We may have to relabel the coordinate functions z_i to do this, but $\sum_{i=1}^{N} dz_i d\overline{z}_i$ is impervious to this relabelling. The reason that one can do this is that if $t_1 = u^{n_1} v^{m_1}$ and $t_2 = u^{n_2} v^{m_2}$ are any two monomials with $n_1 > m_1$, $m_1 < m_2$, the number $a(\Delta) = (n_1-n_2)(m_1-m_2)$ associated to the array $\Delta = \begin{pmatrix} n_1 & m_1 \\ n_2 & m_2 \end{pmatrix}$ is negative. A (u,v)-planar blow-up converts t_1 and t_2 to

$u^{n_1+m_1}v^{m_1}$, $u^{n_2+m_2}v^{m_2}$, resp. in one of the new charts, and $u^{n_1}v^{m_1+n_1}$, $u^{n_2}v^{m_2+n_2}$, resp. in the other chart. The new arrays are therefore $\Delta' = \begin{pmatrix} n_1+m_1 & m_1 \\ n_2+m_2 & m_2 \end{pmatrix}$ and $\Delta'' = \begin{pmatrix} n_1 & m_1+n_1 \\ n_2 & m_2+n_2 \end{pmatrix}$ obtained by the two column operations on Δ. It is easy to check that if $a(\Delta) < 0$ then $a(\Delta')$ and $a(\Delta'')$ are $\geq a(\Delta)+1$. Hence after finitely many steps t_1 and t_2 will give rise to arrays Δ with $a(\Delta) \geq 0$ in <u>all</u> charts. So if $t_1 = u^{n_1}v^{m_1}$, $t_2 = u^{n_2}v^{m_2}$ in any of these new (u,v)-charts, $n_1 > n_2 \Longrightarrow m_1 \geq m_2$ and $n_1 < n_2 \Longrightarrow m_1 \leq m_2$. See [4] for the details.

In fact, we may take a suitable holomorphic root of h_1, a local unit, (where $z_i = u^{n_i}v^{m_i}h_i$, $n_i \geq n_1$, $m_i \geq m_1$) so that we have

$$\begin{aligned} z_1 &= u^{n_1}v^{m_1} \\ z_i &= u^{n_i m_i}h_i \qquad (i = 1,\ldots,N,\ h_i \text{ local units}). \\ &\vdots \\ z_N &= u^{n_N m_N}h_N \end{aligned}$$

Now one decomposes each z_i as a sum

$$z_i = z_{i,1} + z_{i,2}$$

where $z_{i,1}$ = sum of all monomials $a^i_{nm}u^n v^m$ in the power series for z_i with $\det\begin{pmatrix} n_1 & m_1 \\ n & m \end{pmatrix} = 0$

$z_{i,2}$ = sum of all monomials $a^i_{nm}u^n v^m$ in the power series for z_i with $\det\begin{pmatrix} n_1 & m_1 \\ n & m \end{pmatrix} \neq 0$.

Hence $z_{i,1} = \Sigma\alpha^i_j z_1^{\varepsilon_j}$ where $\varepsilon_j \geq 1$ and rational (formally). Note that since planar (u,v) blow-ups result in column operations for exponents of monomials, the decomposition is preserved under such blow-ups. Also since z_i has an absolutely convergent power series, the functions $z_{i,1}$ and $z_{i,2}$ are holomorphic power series in u and v. Also, all the $z_{i,2}$'s cannot be 0, for this would imply that all dz_i's are a multiples of dz_1 in the entire neighborhood, contradicting that the resolution map is a biholomorphism outside (u = 0) or (u=0) ∪ (v=0). Now the same arguments can be applied to the $z_{i,2}$'s to get

$$z_{i,2} = u^{n_i'}v^{m_i'}g_i(u,v) \qquad (i = 1,\ldots,N)$$

where $\det\begin{pmatrix} n_1 & m_1 \\ n_i' & m_i' \end{pmatrix} \neq 0$ and g_i's are 0 or local units, $g_2 \neq 0$ and $n_i' \geq n_2'$ $m_i' \geq m_2'$, (were $z_{2,2} \neq 0$, say, without loss of generality, after a relabelling of 2,...,N). In fact, since g_2 is a non-zero local unit, we may again extract holomorphic roots, etc., so as to keep $u^{n_1}v^{m_1}$ intact and make $u^{n_2'}v^{m_2'}g_2 = u^{n_2'}v^{m_2'}$ (since $n_1m_2'-m_1n_2' \neq 0$) (where the new u and v are the old ones multiplied with local units). To sum up, we have the

Lemma 2.11. After enough blow-ups, one may cover the $\pi^{-1}(V_\varepsilon)$ with (u,v)-neighborhoods so that (up to a reordering of z_i's) we have, in each such neighborhood

$$z_1 = u^{n_1}v^{m_1}$$
$$z_2 = f_2(z_1) + u^{n_2'}v^{m_2'}$$
$$z_i = f_i(z_1) + u^{n_i'}v^{m_i'}g_i, \qquad i = 3,\dots,N$$

where $f_i = \Sigma\alpha_{ij}z_1^{\varepsilon_j}$, $\varepsilon_j \geq 1$ and rational, $g_i = 0$ or a local unit for $i \geq 3$, $n_i' \geq n_2' \geq n_1$, $m_i' \geq m_2' \geq m_1$. (The f_i's are nothing but $z_{i,1}$'s expressed as (formal) functions of z_1. It can be checked that the formal derivatives $\frac{df_i}{dz_1}$ are also holomorphic functions of u and v. Then the singular divisor $\bigcup_i D_i$ is given in this chart by $u = 0$ (if $m_1 = 0$) or ($u=0$, $v=0$) if ($m_1 \neq 0$) corresponding to the cases away from and at a normal crossing, respectively.

2.2 Reduction of the metric

If we pull back the metric $\sum_{i=1}^{N} dz_i\, d\bar{z}_i$ under the coordinate functions to a typical (u,v)-coordinate patch, we have the

Lemma 2.21. With the parametrization of (2.11), the induced metric in a small (u,v)-neighborhood is quasi-isometric to the pullback of $d\zeta_1 d\bar{\zeta}_1 + d\zeta_2 d\bar{\zeta}_2$ under $\zeta_1 = u^{n_1}v^{m_1}$, $\zeta_2 = u^{n_2'}v^{m_2'}$, whenever the latter metric is nonsingular (i.e., outside the singular divisor $\bigcup_i D_i$, which means, $u \neq 0$ if $m_1 = 0$, or $u \neq 0$, $v \neq 0$ if $m_1 \neq 0$).

Proof. Clearly $dz_1 = d\zeta_1$, $dz_2 = (\frac{\partial f_2}{\partial z_1})dz_1 + d\zeta_2 = p_2 d\zeta_1 + d\zeta_2$, where p_2 is a holomorphic function of u and v as remarked at the end of (2.11). Also, for each monomial $u^n v^m$ occurring in $u^{n_i'}v^{m_i'}g_i$, since $\det\begin{pmatrix} n_1 & m_1 \\ n_2' & m_2' \end{pmatrix} \neq 0$, $u^n v^m = \zeta_1^{\alpha}\zeta_2^{\beta}$, where $\alpha,\beta \neq 0$ and by logarithmic differentiation

$$\frac{d(u^n v^m)}{u^n v^m} = \alpha\frac{d\zeta_1}{\zeta_1} + \beta\frac{d\zeta_2}{\zeta_2}, \quad \text{so that}$$

$$d(u^n v^m) = \alpha(\frac{u^n v^m}{\zeta_1})d\zeta_1 + \beta(\frac{u^n v^m}{\zeta_2})d\zeta_2 = \alpha u^{n-n_1}v^{m-m_1}d\zeta_1 + \beta u^{n-n_2'}v^{m-m_2'}d\zeta_2$$

but $n \geq n_i' \geq n_2' \geq n_1$ and $m \geq m_i' \geq m_2' \geq m_1$ so the coefficients of

$d\zeta_1$ and $d\zeta_2$ on the right are holomorphic, i.e., to sum up,

$$dz_i = p_i d\zeta_1 + q_i d\zeta_2 \quad \text{for} \quad i \geq 3$$

where $p_i = \frac{df_i}{dz_1}$ = holomorphic on the neighborhood by (2.11), and

q_i (if $\neq 0$) is also holomorphic by the above. Hence by the Schwarz inequality $\frac{1}{2}|a|^2-|b|^2 \leq |a+b|^2 \leq 2|a|^2 + 2|b|^2$, it follows that

$$dz_1 d\bar{z}_1 = d\zeta_1 d\bar{\zeta}_1$$

$$\varepsilon(\tfrac{1}{2}d\zeta_2 d\bar{\zeta}_2 - |p_2|^2 d\zeta_1 d\bar{\zeta}_1) \leq dz_2 d\bar{z}_2 \leq 2(|p_2|^2 d\zeta_1 d\bar{\zeta}_1 + d\zeta_2 d\bar{\zeta}_2)$$

$$0 \leq dz_i d\bar{z}_i \leq 2\,(|p_i|^2 d\zeta_1 d\bar{\zeta}_1 + |q_i|^2 d\zeta_2 d\bar{\zeta}_2) \quad \text{for} \quad i \geq 3$$

for any ($\varepsilon < 1$) so that

$$(1-\varepsilon|p_2|^2)d\zeta_1 d\bar{\zeta}_1 + \tfrac{\varepsilon}{2} d\zeta_2 d\bar{\zeta}_2 \leq \sum_{i=1}^{N} dz_i d\bar{z}_i \leq (1+2\sum_{i=2}^{N}|p_i|^2)d\zeta_1 d\bar{\zeta}_1 + 2(1+\sum_{i=3}^{N}|q_i|^2)d\zeta_2 d\bar{\zeta}_2 .$$

Since p_i and q_i are holomorphic in the neighborhood, and hence bounded, and also $\varepsilon < 1$ can be chosen a priori so that $\varepsilon|p_2|^2 < \frac{1}{2}$ all over the neighborhood, we get

$$\frac{1}{c}(d\zeta_1 d\bar{\zeta}_1 + d\zeta_2 d\bar{\zeta}_2) \leq \sum_{i=1}^{N} dz_i d\bar{z}_i \leq c(d\zeta_1 d\bar{\zeta}_1 + d\zeta_2 d\bar{\zeta}_2)$$

(for some constant c) all over the (u,v)-neighborhood, proving our claim. #

(2.22) Definition. On a region of the type $B^2-\{0\}\times B^2$ (resp. $B^2-\{0\}\times B^2-\{0\}$) (or a subregion thereof), the metric

$$d\sigma^2 + \sigma^2 d\theta^2 + \sigma^{2c}(dx^2+dy^2)$$

$$(\text{resp. } d\sigma^2 + \sigma^2 d\theta^2 + \sigma^{2c}(d\tau^2+\tau^2 d\phi^2)) \qquad c \geq 1, \text{ rational}$$

(B^2 is the open disc with polar coordinates (σ,θ), and B^2 or $B^2-\{0\}$ in the second factor has the usual coordinates (x,y) or polar coordinates (τ,ϕ) resp.) is called a metric of Cheeger-type.

If we look at the metric $d\zeta_1 d\bar{\zeta}_1 + d\zeta_2 d\bar{\zeta}_2$ as obtained in (2.21) (with $\zeta_1 = u^{n_1}v^{m_1}$, $\zeta_2 = u^{n_2}v^{m_2}$, $n_2 \geq n_1$, $m_2 \geq m_1$, $d = n_1 m_2 - n_2 m_1 > 0$, say) it turns out to be quasi-isometric to a metric of Cheeger type with $c = \frac{n_2}{n_1} \geq 1$. This is not hard to see, if we write $\zeta_2 = \zeta_1^c v^b$ (formally), with $c = \frac{n_2}{n_1}$, $b = m_2 - cm_1 = m_2 - \frac{n_2}{n_1} m_1 = \frac{d}{n_1} > 0$) and differentiate formally to get $d\zeta_2 = c\zeta_1^{c-1}v^b d\zeta_1 + b\zeta_1^c v^{b-1}dv$. Now one plugs this into $d\zeta_1 d\bar{\zeta}_1 + d\zeta_2 d\bar{\zeta}_2$ and uses the Schwarz inequality

$\bar{\alpha}\beta + \alpha\bar{\beta} \leq \varepsilon^2|\alpha|^2 + \frac{1}{\varepsilon^2}|\beta|^2$ to get rid of the cross-terms. v^b can then be written as $x+iy$ or $\tau e^{i\phi}$, and the metric assumes the shape stated. In case $m_1 = 0$, (away from a normal crossing) we get the whole region $B^2-\{0\}\times B^2$ and if $m_1 \neq 0$, (at a normal crossing) we get the subregion of $B^2-\{0\}\times B^2-\{0\}$ defined by $\tau^{m_1/b} > \sigma$ coming from the restriction that $|u| < 1$. Of course if $m_1 = 0$ the condition is vacuous. Also the local variable $\sigma = |\zeta_1|$ can be replaced by the global distance function from the origin $r = (\sum_{i=1}^{N} |z_i|^2)^{1/2}$ up to a quasi-isometry. Indeed the local fields $\frac{\partial}{\partial\sigma}$ piece together to give a global piecewise smooth field ξ which is smooth off codim-1 submanifolds which are defined by the conditions $\tau^{m_1/b} > \sigma$, and subdivide $\pi^{-1}(V_\varepsilon)$ to regions which correspond to the (u,v) neighborhoods obtained in the above discussions. The $B^2-\{0\}\times B^2$ regions correspond to neighborhoods away from normal crossings of $\bigcup_i D_i$ and the subregions of $B^2-\{0\}\times B^2-\{0\}$ correspond to neighborhoods at normal crossings. Precisely, we have the

Lemma 2.23. There exists a piecewise smooth vector field ξ (gradient-like for the function r) such that the associated diffeomorphism

$$f: \quad (0,1) \times N \longrightarrow V_\varepsilon$$

satisfies

(a) f is smooth off a finite number of codim-1 submanifolds H_i which are transverse to the slices $\sigma \times N$ $(= N_\sigma)$, and divide $(0,1)\times N$ into regions as described above.

(b) If σ denotes the first variable, the metric pullback of $\sum_{i=1}^{N} dz_i d\bar{z}_i$ under f is quasi-isometric to a metric of Cheeger type. (The rational exponent $c \geq 1$ changes from region to region.)

We refer to [4] for the details. See [6] for the definition of "gradientlike".

3. L^2-analysis and Proof of the Poincaré Lemma

Let $(B^2-\{0\})\times B^2$ carry a metric of Cheeger type, and x denote coordinates on the link N (which is $S^1\times B^2$ here). Then (following the notation of [1]) we have

Lemma 3.11 Let ω be an i-form on the region which is independent of $d\sigma$ ($\sigma e^{i\theta}$ being the polar coordinate on $B^2-\{0\}$) but depending possibly on σ and x. Then

(i) $\|\omega\|^2_{(0,1)} \leq \int_0^1 \sigma^{2(1-i)+1}\|\omega(\sigma)\|^2_1 \, d\sigma$ for $i \leq 1$

(ii) $\|\omega\|^2_{(0,1)} \geq \int_0^1 \sigma^{2(2-i)-1}\|\omega(\sigma)\|^2_1 \, d\sigma$ for $i \geq 2$

(iii) $\exists$ a sequence $\varepsilon_s \longrightarrow 0$ with

$$\|\omega(\varepsilon_s)\|_\sigma \longrightarrow 0 \text{ as } \varepsilon_s \longrightarrow 0 \text{ for any } \sigma > 0 \text{ and } i \geq 2.$$

(iv) $\|\int_\sigma^a \omega\|^2_\sigma \leq K\sigma \|\omega\|^2_{(0,1)}$ for $i = 0$

$\leq K\sigma \, |\log \sigma| \, \|\omega\|^2_{(0,1)}$ for $i = 1$, $a \in (0,1)$ fixed

$\|\int_0^\sigma \omega\|^2_\sigma \leq K\sigma \|\omega\|^2_{(0,1)}$ for $i \geq 2$.

Proof. Straightforward. One decomposes $\omega = \omega_0 + d\theta \wedge \omega$, where ω_0 and ω_1 are independent of $d\theta$. The volume form for a Cheeger-type metric is $\sigma^{2c+1} \, d\sigma \, dV_N$. Then

$$\|\omega\|^2_{(0,1)} = \int_0^1 \sigma^{2c(1-i)+1} \|\omega_0\|^2_1 \, d\sigma + \int_0^1 \sigma^{2c(2-i)-1}\|d\theta \wedge \omega_1\|^2_1 \, d\sigma$$

$$\overset{\text{def}}{=} \int_0^1 \|\omega_0\|^2_\sigma \, d\sigma + \int_0^1 \|d\theta \wedge \omega_1\|^2_\sigma \, d\sigma \,.$$

Now (i) through (iv) follow exactly as in [1]. See [4] for the details. #

Similar estimates hold for the other kind of region $\subset B^2-\{0\}\times B^2-\{0\}$ with Cheeger-type metric.

Now to prove the Poincare lemma, it is convenient to work with the larger complex of L^2-measurable piecewise smooth forms. This does not change L^2-cohomology, see [1]. We recall that a measurable form ϕ on $(0,1)\times N$ is piecewise smooth (with respect to the codim-1 submanifolds H_i of (2.23)) if (a) ϕ is smooth on $(0,1)\times N - \bigcup_i H_i$, and (b) $\forall y \in H_i$, $\exists$ a neighborhood $U(y)$ of y such that $U(y) - \bigcup_i H_i$ has two components U_1 and U_2 with $\phi|U_j$ extending to a smooth form on $\bar{U}_j$ $(j = 1,2)$ with the same tangential component on $U(y) \cap H_i$. Similarly, piecewise smooth forms on V_ε can also be defined with respect to $f(H_i)$. Piecewise smooth forms on $(0,1)\times N$ and V_ε correspond under f^* and $(f^{-1})^*$.

If α is a piecewise smooth i-form on V_ε, one may express $f^*\alpha$ as

$$f^*\alpha = \phi + d\sigma \wedge \omega$$

in each of the regions in $(0,1)\times N$, as divided by the H_i's, where ϕ and ω are independent of $d\sigma$. Consider the homotopy operators

$$K\alpha = \begin{cases} (f^{-1})^* \int_a^\sigma \omega & \text{for } i \le 2 \quad (a \text{ some fixed number in } (0,1)) \\ (f^{-1})^* \int_0^\sigma \omega & \text{for } i \ge 3 \,. \end{cases}$$

Note that even though ω is discontinuous at the H_i's, $K\alpha$ is continuous.

Proof of the Poincaré Lemma.

Let $[\alpha]$ be an L^2-cohomology class in $H^i_{(2)}(V_\varepsilon)$, with measurable piecewise smooth representative α, (then so is $f^*\alpha$, by 2.23, with respect to the metrics of Cheeger type in each region). Hence the operators $\int_a^\sigma \omega$ and $\int_0^\sigma \omega$ obey the L^2-estimates of (3.11), (iv) in each region with the metric of Cheeger type. Since these estimates do not involve any local parameters, they are globally true for $\int_a^\sigma \omega$ and $\int_0^\sigma \omega$. Hence $K\alpha$ defined above are L^2-norm bounded operators on measurable piecewise smooth forms. Also since (see [1]) in each region the operators $\int_a^\sigma \omega$ and $\int_0^\sigma \omega$ homotope $f^*\alpha$ to $f^*\alpha(a) = \phi(a)$ for $i \le 1$ and 0 for $i \ge 2$ respectively. (This depends on (3.11), (iii) as well), $K\alpha$ homotopes α to $\alpha(a)$ for $i \le 1$ and 0 for $i \ge 2$ respectively. This shows that the map

$$H^i_{(2)}(V_\varepsilon) \longrightarrow H^i_{(2)}(N)$$
$$\alpha \longmapsto \alpha(a) \qquad \text{for } i \le 1$$

and the trivial map of $H^i_{(2)}(V_\varepsilon) \longrightarrow 0$ is injective for $i \ge 2$. The surjectivity follows from the estimates (i) and (ii) of (3.11) which are globally true for $f^*\alpha$ because of involving only the global variable σ. Those estimates show that for $i = 0, 1$, the pullbacks of L^2-forms on $f(a\times N)$ under radial projection are L^2-forms on V_ε. Indeed those estimates also show that both K_α and dK_α are L^2-forms, from the homotopy formula, so that K indeed takes dom d to dom d (where dom d has been enlarged to include L^2-measurable piecewise smooth forms whose distributional derivatives are also L^2). Refer to [1] and [4] for details. The Poincaré lemma is thus established.

4. Generalizations

4.1 Arbitrary surfaces

If V is a surface with a 1-dimensional set of singularities S, we consider a normalization V_1 of V, so that S pulls up to a curve S_1 passing through the singular points $\{p_j\}$ of V_1. Note that $V_1 - S_1 \cong V - S$, so $V-S$ may be equipped with a Fubini metric from a projective embedding of V_1. If the normal V_1 were handled as above, one could arrange that after enough blow-ups the inverse image of S_1 under a resolution of singularities lies in charts away from normal crossings of $\pi^{-1}(p_j) = \bigcup_i D_i$ (where p_j is a typical singular point on V_1). In a typical chart around the point of intersection of S_1 and $\pi^{-1}(p_j)$, the metric was seen to be $d\sigma^2 + \sigma^2 d\theta^2 + \sigma^{2c}(dx^2+dy^2)$ on $B^2-\{0\}\times B^2$, and $\pi^{-1}(S_1 \cap V_1(\varepsilon))$ is, locally, the slice $B^2-\{0\}\times\{0\}$. Removing this set of measure 0 (across which the metric is extending non-singularly, since $S_1 - \{p_1,\ldots,p_\ell\} \subset$ (smooth part of V_1), and the Fubini metric is nonsingular thereon) does not change the analysis at all. Only the links N become noncompact for points on the 0-stratum. At a generic point on S_1, the metric is (non-singular)×(conical), so small neighborhoods of those points are already covered by Cheeger's analysis in [1].

4.2 Threefolds

The analysis of threefolds with isolated singularity, though technically more involved, can be carried through by repeated blow-ups of curves and points after a resolution of singularities. The simplest local models of the metric are

$$d\sigma^2 + \sigma^2 d\theta^2 + \sigma^{2c_1}(dx_1^2 + dy_1^2) + \sigma^{2c_2}(dx_2^2 + dy_2^2)$$

where $c_1, c_2 \geq 1$, rational, and $(\theta_1\ x_i, y_i)$ are local coordinates of the link in a small region. The c_1's and c_2's vary from region to region, as in the surface case.

If there exists a resolution such that $c_1 = c_2$ in all the regions, then L^2-analysis a la Cheeger goes through, and one has L^2-cohomology dual to middle perversity IH_*. This restriction, though rather artificial and hard to verify, does turn out to hold for certain Brieskorn singularities (e.g., $z_1^2+z_2^2 = z_3^3+z_4^3$, see [7]) not earlier covered by Cheeger in [1]. If $c_1 \neq c_2$ in a region, then the estimates fail in an essential way, and a different tack would be needed to prove the conjecture, if true. See [7] for discussion and details.

References

1. Cheeger, J.: On the Hodge Theory of Riemannian pseudomanifolds. Proc. Symp. Pure Math. AMS 36, 91-146 (1980)

2. Cheeger, J., Goresky, M., MacPherson, R.: L^2-cohomology and intersection homology for singular algebraic varieties. Proc. of Year in Differential Geometry, I.A.S., Yau (ed.) Ann. Math. Studies 102, 303-340 (1982)

3. Goresky, M., MacPherson, R.: Intersection homology theory. Topology 19, 135-162 (1980). Intersection homology II, Invent. Math. 72, 77-130, (1983)

4. Hsiang, W-C., Pati, V.: L^2-cohomology of normal algebraic surfaces, Invent. Math. 81, 395-412 (1985)

5. Laufer, H.: Normal Two-dimensional Singularities, Ann. Math. Stud., Princeton University Press (1971)

6. Milnor, J.: Lectures on the h-cobordism theorem (Notes by L. Siebenmann and J. Sondow.) Princeton University Press (1965)

7. Pati, V.: L^2-Cohomology of Algebraic Varieties, Ph.D. thesis, Princeton (1985)

SOME REMARKS ABOUT THE HODGE CONJECTURE

J.H.M. Steenbrink *)
Department of Mathematics
University of Leiden
Niels Bohrweg 1
2333 CA Leiden
The Netherlands.

*) Supported by the SLOAN foundation.

In this paper we summarize a few facts, related to the Hodge conjecture, which we fell upon preparing talks about this subject at Oberwolfach (1981) and Utrecht (1985). We do not aim to give a survey of known results. For this we refer the reader to the excellent paper of Shioda [13].

I thank J.P. Murre for valuable discussions. This paper was written while the author was a visitor of the Institute for Advanced Study at Princeton U.S.A.. We thank it for its hospitality.

§1. GROTHENDIECK'S FORMULATION OF THE HODGE CONJECTURE

Let $m \in \mathbb{Z}$. A $\mathbb{Q}$-*Hodge structure* of weight m is a pair $(V_{\mathbb{Q}}, F^{\cdot})$ consisting of a finite-dimensional $\mathbb{Q}$-vector space $V_{\mathbb{Q}}$ together with a decreasing filtration $F^{\cdot}$ on $V_{\mathbb{C}} = V_{\mathbb{Q}} \otimes \mathbb{C}$ (the *Hodge filtration*) such that $V_{\mathbb{C}} = F^p \oplus \bar{F}^q$ whenever $p+q = m+1$. It is assumed that $F^p = V_{\mathbb{C}}$ for $p << 0$ and $F^p = 0$ for $p >> 0$.

Let X be a smooth complex projective variety. Then for $m \in \mathbb{N}$, the cohomology groups $H^m(X,\mathbb{Q})$ are $\mathbb{Q}$-Hodge structures of weight m. The subspace F^p of $H^m(X,\mathbb{C})$ can be described as the cohomology classes which are represented by harmonic forms which locally contain at least p dz's.

For every $\mathbb{Q}$-Hodge structure $(V_{\mathbb{Q}}, F^{\cdot})$ of weight m one has its *Hodge decomposition*

$$V_{\mathbb{C}} = \bigoplus_{p+q=m} V^{p,q}$$

with $V^{p,q} = F^p \cap \overline{F^q}$. We will only consider Hodge structures with

$V^{p,q} \neq 0 \Rightarrow p,q \geq 0$, i.e. with $F^0 = V_{\mathbb{C}}$.

The *level* of a non-zero $\mathbb{Q}$-Hodge structure is $\max\{|p-q| : V^{p,q} \neq 0\}$. For the Hodge structures we consider, the level will always be less than or equal to the weight m , with equality if and only if $V^{m,0} \neq 0$.

Let Z be an algebraic subvariety of codimension p in a smooth projective variety X of dimension n . Then, dualizing the restriction map $H^{2n-m}(X) \to H^{2n-m}(Z)$ with respect to cup product, we obtain the map

$$\gamma: H^m_Z(X) \to H^m(X)$$

which fits in the exact sequence

$$H^m_Z(X) \overset{\gamma}{\to} H^m(X) \xrightarrow{i^*} H^m(X \setminus Z) .$$

If $\tilde{Z} \to Z$ is a resolution of singularities, the sequence above gives rise to

$$H^{m-2p}(\tilde{Z}) \overset{\tilde{\gamma}}{\to} H^m(X) \xrightarrow{i^*} H^m(X \setminus Z) ,$$

in which $\tilde{\gamma}$ (the *Gysin map*) is a morphism of Hodge structures of type (p,p) . We obtain from [6,(8.2.8.1)] that $\operatorname{Ker} i^* = \operatorname{Im} \tilde{\gamma}$; in particular one obtains

(1.1) THEOREM. *If* $Z \subset X$ *is a codimension* p *subvariety in a smooth projective variety* X , *then* $\ker(i^*: H^m(X,\mathbb{Q}) \to H^m(X \setminus Z, \mathbb{Q}))$ *is a* $\mathbb{Q}$*-Hodge substructure of* $H^m(X,\mathbb{Q})$ *of level* $\leq m-2p$. □

The general Hodge conjecture, in Grothendieck's improved formulation ([9]) is, that for all X,m,p :
GHC(X,m,p): "*For every* $\mathbb{Q}$*-Hodge substructure* V *of* $H^m(X,\mathbb{Q})$ *with level* $\leq m-2p$ *there exists a subvariety* Z *of* X *of codimension* p , *such that* $V \subset \ker(H^m(X,\mathbb{Q}) \to H^m(X \setminus Z, \mathbb{Q}))$".

A special case of this is the so-called (p,p)-conjecture GHC$(X,2p,p)$ which states that every element of $H^{2p}(X,\mathbb{Q}) \cap H^{p,p}$ should be a rational multiple of the cohomology class of an algebraic cycle.

REMARK. Our GHC(X,m,p) is G-Hodge (X, F^pH^m) in Shioda's notation.

§2. GROTHENDIECK'S INDUCTION PRINCIPLE

We will formulate and sketch a proof of an inductive argument of Grothendieck; although it forces one to work with the general Hodge conjecture, we will show that in certain cases it enables one to conclude about the usual Hodge conjecture too, by using a trivial trick:

(2.1) LEMMA. *Let* X *be a smooth projective variety. Then if* $GHC(X,2p,p-1)$ *holds,* $GHC(X,2p,p)$ *also holds.*

Proof. Suppose $V \subset H^{2p}(X,\mathbb{Q})$ is a $\mathbb{Q}$-Hodge substructure of level 0, i.e. purely of type (p,p). Then V has level ≤ 1, so by $GHC(X,2p,p-1)$ there exists $Z \subset X$ of codimension $p-1$ such that V is in the image of Gysin map $\tilde{\gamma}: H^2(\tilde{Z}) \to H^{2p}(X)$. ($\tilde{Z} \to Z$ is a resolution of singularities as before). Now $\mathrm{Ker}(\tilde{\gamma}) \subset H^2(\tilde{Z})$ has a complementary subspace which is a $\mathbb{Q}$-Hodge substructure (because $H^2(\tilde{Z})$ is a polarizable Hodge structure). Our Hodge structure V corresponds therefore to a $\mathbb{Q}$-Hodge substructure of type $(1,1)$ of $H^2(\tilde{Z})$, which by Lefschetz' theorem is supported by a divisor $\tilde{D}$ on $\tilde{Z}$. Let D be the image of $\tilde{D}$ under $\tilde{Z} \to Z \to X$. Then $V \subset \ker(H^{2p}(X) \to H^{2p}(X \setminus D))$ and D has codimension p in X. □

(2.2) THEOREM. (Grothendieck [9]) *Let* X *be a smooth* n*-dimensional projective variety. Let* Y *be a general hyperplane section of* X. *Suppose that* $GHC(Y,n-2,p-1)$ *holds, and that there exists a subvariety* $Z \subset Y$ *of codimension* p *such that the orthogonal complement* E *of* $H^{n-1}(X)$ *in* $H^{n-1}(Y)$ *maps to zero under* $H^{n-1}(Y) \to H^{n-1}(Y \setminus Z)$. *Then* $GHC(X,n,p)$ *holds.*

Before sketching the proof of this theorem, we consider some applications.

1. Let X_3^3 be a cubic threefold in $\mathbb{P}^4$. Then Y is a cubic surface, so $H^2(Y)$ has level zero and is supported by a divisor on Y. Hence all of $H^3(X)$, which has level 1, is supported by a divisor on X, i.e. $GHC(X,3,1)$ holds.
2. Let X_3^4 be a cubic threefold in $\mathbb{P}^5$. Its hyperplane section is an X_3^3, so $GHC(X_3^4,4,1)$ holds by Theorem (2.2). Hence $GHC(X_3^4,4,2)$ also holds by Lemma (2.1). (This has been proved by Griffiths, Zucker [14] and Murre [12] by totally different means.)

To get more examples, we consider Abel-Jacobi-like mappings. Let X be smooth projective of dimension n, and let $(Z_s)_{s\in S}$ be a family of algebraic cycles on X of codimension p, parametrized by a smooth projective variety S (e.g. a resolution of a component of the Chow variety of X). Then for each k with $2p \leq k \leq \dim S+2p$ we have the "tube-over-a-cycle map"

$$\tau_k: H_{k-2p}(S) \to H_k(X) .$$

(2.3) LEMMA. *Suppose that* τ_k *is surjective. Then* $H^{2n-k}(X)$ *has level* $\leq k-2p$ *and* $GHC(X,2n-k,n-k+p)$ *holds.*

Proof. In view of the weak Lefschetz theorem on hyperplane sections we may assume that $\dim(S) = k-2p$. Let $Z = \bigcup_{s\in S} Z_s \subset X$. Then it follows that the map

$$H_Z^{2n-k}(X) \to H^{2n-k}(X)$$

is surjective. Moreover Z has dimension $\leq k-2p+p = k-p$, so $\operatorname{codim}(Z,X) \geq n-k+p$. □

There are numerous examples of surjective Abel-Jacobi mappings (often shown to be isomorphisms). E.g. for conics on the quartic threefold (Letizia [11]); hence $GHC(X,4,1)$ and $GHC(X,4,2)$ hold for the quartic fourfold in $\mathbb{P}^5$; or for $\mathbb{P}^2$'s on a cubic fivefold (Collino), hence $GHC(X,6,2)$ and $GHC(X,6,3)$ hold. Similar arguments prove the Hodge conjecture for even-dimensional intersections of two or three quadrics.

Before pssing to the proof of (2.2), we need two more lemmas.

(2.4) LEMMA. *Let* X *be a smooth projective variety, defined over a subfield* k *of* $\mathbb{C}$. *Let* $V \subset H^n(X,\mathbb{Q})$ *be a Hodge substructure, and* $Z \subset X$ *a subvariety of codimension* p *such that* $V \subset \operatorname{Ker}(H^n(X) \to H^n(X\setminus Z))$. *Then there exists a subvariety* Z_0 *of* X , *defined over* k , *with* $V \subset \operatorname{Ker}(H^n(X) \to H^n(X\setminus Z_0))$.

Proof. Z is defined over a finitely generated extension K of k . It suffices to consider the cases where K/k is purely transcendental or finite Galois. In the latter case we let Z_0 be the union of all conjugates of Z over k . If K/k is purely transcendental of degree m , the equations of Z define a k-subvariety $\tilde{Z} \subset X \times \mathbb{A}^m$; the image of $R^n_{\tilde{Z}}(p_2)_*\mathbb{Q}_{X\times\mathbb{A}^m} \to R^n(p_2)_*\mathbb{Q}_{X\times\mathbb{A}^m}$ is a constructible subsheaf of the constant sheaf on $\mathbb{A}^m$ with fibre $H^n(X,\mathbb{Q})$, hence it is constant on a dense Zariski-open subset U. Choose a rational point $t_0 \in U$ and let $Z_0 = p_2^{-1}(t_0)$. This will do. □

(2.5) LEMMA. *Let* $\bar{f}: \bar{X} \to \bar{S}$ *be a flat morphism of smooth projective varieties,* $\dim \bar{S} = 1$. *Let* $S \overset{j}{\hookrightarrow} \bar{S}$ *be Zariski open and dense such that the induced map* $f: X = \bar{f}^{-1}(S) \to S$ *is smooth. Then* $H^1(\bar{S}, j_*R^{k-1}f_*\mathbb{Q}_X)$ *is a* $\mathbb{Q}$*-Hodge substructure of* $H^k(\bar{X},\mathbb{Q})$.

Proof. The Leray spectral sequence for $\bar{f}$ is a spectral sequence of Hodge structures cf. [15, Theorem (15.16)], hence

$$E_2^{1,k-1} = E_\infty^{1,k-1} = H^1(\bar{S}, R^{k-1}\bar{f}_*\mathbb{Q}_{\bar{X}})$$

is a $\mathbb{Q}$-Hodge substructure of $H^k(\bar{X},\mathbb{Q})$ (use the polarizability to identify subquotients with subspaces). Moreover, by the local invariant cycle theorem one has a surjection

$$R^{k-1}\bar{f}_*\mathbb{Q}_{\bar{X}} \to j_*R^{k-1}f_*\mathbb{Q}_X$$

whose kernel has zero-dimensional support; hence $H^1(\bar{S}, R^{k-1}\bar{f}_*\mathbb{Q}_{\bar{X}}) \tilde{\to}$ $\tilde{\to} H^1(\bar{S}, j_*R^{k-1}f_*\mathbb{Q}_X)$. □

We now sketch the proof of (2.2). Choose a Lefschetz pencil of hyperplance sections $\{Y_t\}_{t\in\mathbb{P}^1}$ of X. Its generic member Y_t is defined over $k(t)$, if k is a field of definition of X. Let E_t be the orthogonal complement of $H^{n-1}(X)$ in $H^{n-1}(Y_t)$. There exists a subvariety Z_t of Y_t of codimension p, defined over $k(t)$, such that the image of

$$H^{n-1}_{Z_t}(Y_t) \to H^{n-1}(Y_t)$$

contains E_t. The only difficult part of $H^n(X)$ to deal with is $H^1(\mathbb{P}^1, j_*E)$ where $U \overset{j}{\hookrightarrow} \mathbb{P}^1$ is Zariski-open such that Y_t is smooth for $t \in U$ and E is the local system on U with fibre E_t. Let $\pi: \tilde{X} \to \mathbb{P}^1$ be the map with fibres Y_t. Then there exists a unique subvariety Z' of $\tilde{X}$, flat over $\mathbb{P}^1$, with fibre $Z_t \subset Y_t$. Claim:

$$H^1(\mathbb{P}^1, j_*E) \subset \mathrm{Image}(H^n_Z(X) \to H^n(X))$$

where Z is the image of Z' in X .

Proof. Let $\tilde{Z} \to Z'$ be a resolution and let $g: \tilde{Z} \to \mathbb{P}^1$ be the induced morphism. We may assume that $U \subset \mathbb{P}^1$ is so small that g is smooth over U . Then the image of

$$j^* R^{n-1-2p} g_* \mathbb{Q}_{\tilde{Z}} \to j^* R^{n-1} \pi_* \mathbb{Q}_X$$

contains E ; because these local systems on U underly polarizable variations of Hodge structure, E is a direct factor of $j^* R^{n-1-2p} g_* \mathbb{Q}_{\tilde{Z}}$ ([5],Théorème (4.2.6)). Hence $H^1(\mathbb{P}^1, R^{n-1-2p} g_* \mathbb{Q}_{\tilde{Z}})$ contains $H^1(\mathbb{P}^1, j_* E)$ as a direct factor. □

REMARK. In the preceding examples, one needed specific information about algebraic cycles on X , i.e. concerning the Abel-Jacobi map. A lot of cases can be done with the following

(2.6) PROPOSITION. *Let* X *be a smooth* n*-dimensional projective variety. Suppose* X *is covered by rational curves* (X *is "uniruled"). Then* GHC(X,n,1) *holds.*

The proof is similar to the case $n = 4$, due to Conte and Murre [3]. This applies in particular to varieties X for which $-K_X$ is numerically effective and $K_X^n \neq 0$ because these are uniruled [10].

§3. SEMIREGULARITY OF CERTAIN SUBSCHEMES OF HYPERSURFACES

In [8,footnote 13], Grothendieck stated a conjecture which is weaker than the Hodge (p,p) conjecture:

(VHC) *Suppose that* $f: X \to S$ *is a smooth projective morphism with* S *connected, smooth. Suppose that* $\lambda \in H^0(S, R^{2p} f_* \mathbb{Q}_X)$ *is of type* (p,p) *everywhere, and for some* $s_0 \in S$, $\lambda(s_0)$ *is the cohomology class of an algebraic cycle of codimension* p *on* X_{s_0} . *Then* $\lambda(s)$ *is an algebraic cycle class for all* $s \in S$.

This "variational Hodge conjecture" may be attacked by comparison of the following two deformation functors. Let $\mathcal{C}$ denote the category of local Artinian $\mathbb{C}$-algebras with residue class field $\mathbb{C}$. For X a smooth projective variety over $\mathbb{C}$ and $Z \subset X$ a subscheme define $D(X,Z)$ the functor from $\mathcal{C}$ to sets with $D(X,Z)(A)$ = the set of isomorphism classes of diagrams

$$\begin{array}{ccccc} Z_0 & \hookrightarrow & X_0 & \hookrightarrow & \mathrm{Spec}(\mathbb{C}) \\ \downarrow & & \downarrow & & \downarrow \\ Z & \hookrightarrow & X & \twoheadrightarrow & \mathrm{Spec}(A) \end{array}$$

with X smooth over A, Z flat over A and both squares Cartesian. If $X \to \mathrm{Spec}(A)$ is an infinitesimal deformation of X, the relative De Rham cohomology group $H^n_{DR}(X/A)$ is a free A-module of rank $b_n(X)$ ([1],Prop.(3.8)) and there is a Gauss-Manin connection.

$$\nabla: H^n_{DR}(X/A) \to \Omega^1_{A/\mathbb{C}} \otimes_A H^n_{DR}(X/A) .$$

An element $\sigma \in H^n_{DR}(X/A)$ is called *horizontal* if $\nabla\sigma = 0$. Moreover $H^n_{DR}(X/A)$ carries a Hodge filtration $F^{\cdot}$, and each $F^pH^n_{DR}(X/A)$ is a free A-submodule. Every $\lambda \in H^n(X,\mathbb{C})$ has a unique extension $\lambda_A \in H^n_{DR}(X/A)$ such that $\lambda_A \otimes \mathbb{C} = \lambda$ and $\nabla(\lambda_A) = 0$: its *"horizontal extension"*.

Let X,Z be as before with $Z \subset X$ of pure codimension p. Let λ be the cohomology class of Z. Then we have the functor

$$D(X,\lambda): C \longrightarrow \mathrm{Sets} ,$$

$D(X,\lambda)(A)$ = the set of isomorphism classes of Cartesion diagrams

$$\begin{array}{ccc} X & \to & \mathrm{Spec}(\mathbb{C}) \\ \downarrow & & \downarrow \\ X & \to & \mathrm{Spec}(A) \end{array}$$

such that the horizontal extension λ_A of λ lies in $F^pH^{2p}_{DR}(X/A)$. Then clearly $D(X,\lambda)$ is a subfunctor of $\mathrm{Def}(X)$. The forgetful morphism $D(X,Z) \to \mathrm{Def}(X)$ factorizes through $D(X,\lambda)$ as for each $(Z \hookrightarrow X \to \mathrm{Spec}(A)) \in D(X,Z)(A)$, there exists a relative cycle class in $H^{2p}_{DR}(X/A)$, which is a (the) horizontal extension of λ and lies in F^p (cf. [1],§5).

If for given (X,Z), the morphism

$$c\ell: D(X,Z) \to D(X,\lambda)$$

is *smooth*, i.e. for each surjection $A' \to A$ in C the map

$D(X,Z)(A') \to D(X,\lambda)(A)$ is surjective, then each deformation of X which preserves λ as a Hodge class can be followed by a deformation of Z.

Bloch [1] gave a sufficient condition for $c\ell$ to be smooth; if Z is locally a complete intersection, there is a natural map

$$u\colon H^1(N_{Z/X}) \to H^{p+1}(X,\Omega_X^{p-1})$$

(the *semiregularity map*). If u is injective, Z is called *semiregular in* X. This implies smoothness of the functor $c\ell$ and hence of its fibre, the completion of Hilb(X) at the point corresponding to Z ([1],Theorem (7.3)). We consider the case where X is a smooth hypersurface in $\mathbb{P}^{2m+1}$, given by a homogeneous polynomial $f \in S = \mathbb{C}[x_0,\dots,x_{2m+1}]$ of degree d, and $Z \subset X$ is a subvariety of codimension m, which is a complete intersection in $\mathbb{P} = \mathbb{P}^{2m+1}$. Let $\xi_0,\dots,\xi_m$ be homogeneous generators of $I(Z) \subset S$, with $d_i = \deg \xi_i$.

(3.1) THEOREM. *With these notations,* Z *is semiregular in* X.

The only fact we need to know about the semiregularity map is, that it fits in a commutative diagram

$$\begin{array}{ccc} H^1(X,\Theta_X) & \xrightarrow{\cup\lambda} & H^{m+1}(X,\Omega_X^{m-1}) \\ \downarrow & & \uparrow u \\ H^1(Z,\Theta_X \otimes \mathcal{O}_Z) & \xrightarrow{\alpha} & H^1(Z,N_{Z/X}) \end{array}$$

where λ is the cohomology class of Z and α is induced by the surjection $\Theta_X \otimes \mathcal{O}_Z \to N_{Z/X}$ (cf. [1],Prop. (6.2)). Because $Z \subset X$, we can write

$$f = \sum_{i=0}^{m} \xi_i \eta_i$$

with either $\eta_i = 0$ or η_i is homogeneous of degree $d-d_i$. Observe that $\xi_0,\dots,\xi_m,\eta_0,\dots,\eta_m$ do not have a common zero, as this would give a singular point of X. So either one of the η_i, say η_0, is a unit, or $(\xi_0,\dots,\xi_m,\eta_0,\dots,\eta_m)$ is a regular sequence in S, generating an ideal J of finite codimension. If η_0 is a constant, we may as well take $\xi_0 = f$, and λ will be a multiple of the class of $X \cap \mathbb{P}^{m+1}$. Then, if $m > 1$, $H^1(N_{Z/X}) \cong \bigoplus_{i=1}^{m} H^1(\mathcal{O}_Z(d_i)) = 0$ in view of the fact, that $H^i(\mathcal{O}_V(j)) = 0$ for V an m-dimensional complete intersection in projective space, and for all j, as soon as $1 \le i \le m-1$ (easy to prove by induction on m). The case $m = 1$ is easy, and left to the reader.

So assume no η_i is a unit. From the exact sequence

$$\begin{array}{ccccccccc} 0 & \to & N_{Z/X} & \to & N_{Z/\mathbb{P}} & \to & N_{X/\mathbb{P}} \otimes \mathcal{O}_Z & \to & 0 \\ & & & & \wr\wr & & \wr\wr & & \\ & & & & \bigoplus_{i=0}^{m} \mathcal{O}_Z(d_i) & \xrightarrow{\beta} & \mathcal{O}_Z(d) & & \end{array}$$

we obtain an exact sequence (again $m > 1$ for simplicity):

$$\bigoplus_{i=0}^{m} H^0(\mathcal{O}_Z(d_i)) \xrightarrow{\beta} H^0(\mathcal{O}_Z(d)) \xrightarrow{\gamma} H^1(N_{Z/X}) \to 0$$

in which β is given by

$$\beta(u_0,\ldots,u_m) = \sum_{i=0}^{m} u_i\eta_i \pmod{I(Z)} .$$

Because γ is surjective, u is injective iff $\ker u\circ\gamma = \ker\gamma = \operatorname{Im}\beta$. One has a commutative diagram

$$(3.2) \qquad \begin{array}{ccccc} & & S_d & \xrightarrow{U\lambda} & H^{m+1}(X,\Omega_X^{m-1}) \\ & & \downarrow\delta & & \uparrow u \\ \bigoplus_{i=0}^{m} S_{d_i}/I(Z)_{d_i} & \xrightarrow{\beta} & S_d/I(Z)_d & \xrightarrow{\gamma} & H^1(Z,N_{Z/X}) \end{array} .$$

Now the map $U\lambda$ in the upper row can be described explicitly as follows. Let $R = S/(\partial_0 f,\ldots,\partial_{2m+1} f)$. One has isomorphisms

$$R_{\mu+id} \xrightarrow{\sim} H^{m+i}(X,\Omega_X^{m-i})_0$$

($\mu = (m+1)(d-2)$, $i = -m,\ldots,m$, the subscript 0 means primitive cohomology), such that if $A \in R_\mu$ corresponds to the primitive part of λ, the map $S_d \xrightarrow{U\lambda} H^{m+1}(X,\Omega_X^{m-1}) \cong R_{\mu+d}$ is just multiplication by A (up to a constant non-zero factor) (cf. [2],Thm. 3). Remark that A cannot be zero: if $A = 0$, Z would be homologically equivalent to a positive multiple of $X \cap \mathbb{P}^{m+1}$, in contradiction with the fact that Z is disjoint from the subvariety of dimension m, given by $\eta_0 = \ldots = \eta_m = 0$, Again the multiplication map

$$m: R_\mu \times R_\mu \to R_{2\mu} \cong \mathbb{C}$$

is cup product up to a constant factor, so A^2 generates $R_{2\mu}$. Moreover m is a non-degenerate pairing, so $S_\mu \cong \mathbb{C}\cdot A \oplus \mathrm{Ann}(A)_\mu$ where $\mathrm{Ann}(A) = \mathrm{Ker}(S \xrightarrow{\cdot A} R)$. Now observe that $S_d \xrightarrow{\cdot A} R_{\mu+d}$ factorizes as in (3.2). This implies that $\mathrm{Ann}(A)_d$ contains J_d. Hence $\mathrm{Ann}(A)$ contains J_μ, but it is easily checked that μ is the degree of the socle of the 0-dimensional Gorenstein ring S/J, so J_μ has codimension one in S_μ. Hence $J_\mu = \mathrm{Ann}(A)_\mu$, and consequently $\mathrm{Ann}(A) = J$ because they are Gorenstein ideals with the same socle. In particular, $\mathrm{Ann}(A)_d = J_d$ which implies that $\mathrm{Ker}(U\lambda) = \mathrm{Ker}\, u\gamma\delta = \mathrm{Ker}\, \gamma\delta$ so u is injective, as $\gamma\delta$ is surjective. □

COROLLARY OF THE PROOF. *If* $d_i < d/2$ *for all* i, *one can recover* Z *from* λ.

Indeed, λ determines the ideal J, and $I(Z)$ is the ideal in S, generated by $\Sigma_{i<d/2} J_i$ in that case.

REMARKS. 1. Statement and proof of Theorem (3.1) arose from an attempt to understand the computations of Griffiths and Harris in [7], Chapter 4.

2. Together with R-O. Buchweitz, the author is trying to generalize Theorem (3.1) to more general subschemes of hypersurfaces, using maximal Cohen-Macaulay modules on the affine cone of the hypersurface.

3. Here is an explicit formula for A in terms of ξ's an η's: take

$$A = \det\left(\frac{\partial \xi_i}{\partial x_j}, \frac{\partial \eta_k}{\partial x_j}\right) .$$

That this is the right formula can be checked by considering the quadric $Q : \sum_{i=0}^{m} x_i y_i = 0$ in (weighted) projective $(2m+1)$-space and pulling back via the morphism $X \to Q$, given by $x_i = \xi_i$, $y_k = \eta_k$.

REFERENCES

[1] S. BLOCH: *Semi-Regularity and the Rham Cohomology.* Invent. Math. 17, 51-66 (1972).

[2] J. CARLSON & Ph.A. GRIFFITHS: *Infinitesimal variations of Hodge structure and the global Torelli problem.* In: A. Beauville ed.: Algebraic Geometry, Angers 1979, Sijthoff & Noordhoff 1980; pp. 51-76.

[3] A. CONTE & J.P. MURRE: *The Hodge conjecture for fourfolds admitting a covering by rational curves.* Math. Ann. 238, 79-88 (1978).

[4] P. DELIGNE: *Théorème de Lefschetz et critères de dégénérescence de suites spectrales.* Publ. Math. IHES 35, 107-126 (1968).

[5] P. DELIGNE: *Théorie de Hodge II.* Publ. Math. IHES 40, 5-57 (1971).

[6] P. DELIGNE: *Theorie de Hodge III.* Publ. Math. IHES 44, 5-77 (1974).

[7] Ph. GRIFFITHS & J. HARRIS: *Infinitesimal variations of Hodge structure (II): an infinitesmial invariant of Hodge classes.* Compos. Math. 50, 207-265 (1983).

[8] A. GROTHENDIECK: *On the de Rham cohomology of algebraic varieties.* Publ. Math. IHES 29, 95-103 (1966).

[9] A. GROTHENDIECK: *Hodge's general conjecture is false for trivial reasons.* Topology 8, 299-303 (1969).

[10] J. KOLLAR: letter to J.P. Murre.

[11] M. LETIZIA: *The Abel-Jacobi mapping for the quartic threefold.* Invent. Math. 75, 477-492 (1984).

[12] J.P. MURRE: *On the Hodge conjecture for unirational fourfolds.* Indag. Math. 80, 230-232 (1977).

[13] T. SHIODA: *What is known about the Hodge conjecture?* Advanced Studies in Pure Mathematics I, 1983. Algebraic varieties and analytic varieties, pp. 55-68.

[14] S. ZUCKER: *The Hodge conjecture for cubic fourfolds.* Compos. Math. 34, 199-209 (1977).

[15] S. ZUCKER: *Hodge theory with degenerating coefficients:* L_2 *cohomology in the Poincaré metric.* Annals of Math. 109, 415-476 (1979).